SCIENTIFIC MANAGEMENT
IN
ACTION

SCIENTIFIC MANAGEMENT
IN
ACTION

Taylorism at Watertown Arsenal,

1 9 0 8 – 1 9 1 5

By
HUGH G. J. AITKEN

New foreword by
Merritt Roe Smith

PRINCETON UNIVERSITY PRESS

Published by Princeton University Press, 41 William Street,
Princeton, New Jersey 08540
In the United Kingdom. Princeton University Press,
Guildford, Surrey

LCC 84–26462
ISBN 0–691–04241–1
ISBN 0–691–00375–0 (pbk.)

Original title: *Taylorism at Watertown Arsenal. Scientific
Management in Action, 1908–1915*

Clothbound editions of Princeton University Press books are printed
on acid-free paper, and binding materials are chosen for strength
and durability. Paperbacks, while satisfactory for personal collec-
tions, are not usually suitable for library rebinding.

Printed in the United States of America by Princeton University Press,
Princeton, New Jersey

FOREWORD

The social history of technology has become quite fashionable in recent years, but it is not new. In fact, one of the best works in the field actually predates it by more than a decade. I refer, of course, to Hugh G. J. Aitken's masterful study of *Scientific Management in Action*.

Taylorism, the shorthand expression for Frederick W. Taylor's brand of scientific management, is one of the most frequently discussed topics in American business and technological history. General textbook writers often address the subject, and students of the Progressive Era invariably refer to it as an example of the reformist impulse to rationalize American society. In their efforts to make sense of Taylorism and place it in context, historians have brought a variety of interpretative perspectives to the study of scientific management. Some, like Frank B. Copley, Sudhir Kakar, and Daniel Nelson, approach the subject through the medium of biography and provide absorbing accounts of Taylor and his circle.[1] Others, like Samuel Haber, examine the movement through the prism of politics in order to reveal basic changes in the country's political structure at the turn of the twentieth century.[2] Still others view Taylorism as a critical juncture in the rise of modern industrial management and, depending on their point of view, depict the movement either as a force for more efficient business administration or as a frontal assault on the traditional prerogatives of labor.[3] Yet, as good as they are, none of these writers surpass Hugh Aitken in portraying the complex technical and human relationships that comprised scientific management.

When *Scientific Management in Action* first appeared in

1960 under the title *Taylorism at Watertown Arsenal*, it received an enthusiastic reception from reviewers. In his contribution to *The American Historical Review*, Alfred D. Chandler, Jr. hailed the volume as "the clearest picture yet written on the nature and significance of scientific management." Moreover, everyone who reviewed the book reached much the same conclusion. Writing in the *Business History Review*, George S. Gibb praised the book as "extremely competent" and "a classic in its field." John B. Rae agreed. "This book will be indispensable," he observed in the *Journal of Economic History*; "it can well be used as a model by those who believe that scholarly history can be written with literary charm." In fact, the only negative comment came from Gibb who mildly chided Aitken for being too cautious and restrained "in relating the Watertown episode to broad forces at work elsewhere" in American society. Gibb quickly added, however, that his remark "probably is not a criticism at all."[4]

Twenty-five years have elapsed since the book's publication, yet it remains as fresh today as it was in 1960. A question worth asking is why this is so. What accounts for the books's remarkable longevity and influence?

To my mind, three attributes—clarity, insight, and originality—distinguish the volume and give it special standing in the world of scholarship. There is no need to belabor the subject of clarity. As most readers immediately recognize, Aitken excels at describing the Taylor system and relating its component parts. But more than that, he reveals how various actors in the Watertown story—ordnance officers, arsenal workers, and Taylor associates—perceived the changes that were taking place and how they responded to them. The installation of Taylor's system at Watertown turns out to be a complicated saga fraught with misunderstanding, distrust, and conflict. Aitken succeeds in unraveling these complex social relations because he obviously understands the subject and writes with authority and feeling. By couching his study within a well-defined community setting, he also is able to move back and forth between the general and specific aspects of the subject and to use one aspect as a means of illuminating the other. Each chapter thus provides a different angle of vision on scientific management. With each view comes an enhanced understanding of the system's complexity and importance.

Aitken's account is filled with revealing insights into the nature of scientific management. We learn, for example, that the primary goal of Taylorism was not merely to get employees to work harder but to control the entire job situation. We also learn that standardization became the primary means of achieving control. From the planned scheduling and routing of work in progress to the use of uniform belting and high-speed steel cutting tools, all of the innovations that together comprised scientific management were designed, as Aitken puts it, "to achieve total control of the job and its performance and in particular to enable management to prescribe and enforce a standard work pace" (pp. 28–29). Taylor's insistence that "no time studies should be attempted until all working conditions had been brought up to a high level of efficiency" reveals how vital uniform standards were to his system of management. Until Aitken clarified the nexus between standardization and control, few people fully appreciated this important aspect of scientific management.

An underlying premise of Taylorism, one that produced serious discord at Watertown, held that knowledge of every aspect of production was necessary for managerial control. "If Taylor was to attain his primary goal of securing complete control over the pace of work," Aitken writes, "it was essential that he know precisely the maximum rate of output of which his machines were capable" (p. 31). This meant that managers or their surrogates, the efficiency experts, had to enter the shop, observe workers at close range, and scrupulously record what they saw in an effort to acquire an understanding of individual production tasks. Doing so, of course, was touchy business, particularly when the observer entered the shop unannounced with a stop watch in his hand. In delineating the sources of conflict at Watertown, Aitken reveals, as no one had before, that knowledge was power and that workers sensed the threat time studies posed to their autonomy. Such studies were intolerable because they aimed at transferring the skills of the worker to the engineer. Faced with the loss of the one thing that insured their power on the shop floor, the molders at Watertown expressed themselves eloquently about their fear of dispossession. Aitken quotes one molder's statement that "I don't like a man to stand over me with a stop watch because it looks to me as if it is getting down to slavery." Others "felt hustled and driven and resented the idea of being set in competition with each other." It was "humiliating" and "un-American" (pp. 216, 150). When the molders walked out of the shop on August 11, 1911, Taylor knew that "a fundamental issue was at stake." "This strike hits at the very foundation of scientific management," he wrote to his colleague Carl Barth, "and if the owners of the company or the government are not to be allowed to obtain exact information, then scientific management becomes impossible" (p. 164). Taylor realized that technical knowledge was the basis of power and that "no limits could be placed on management's right to know" (p. 165). Thanks to Aitken's penetrating analysis, we now appreciate the full significance time studies had for labor-management relations under the Taylor system.

Other important insights enhance the book's reputation as a scholarly work. Aitken recognizes, for instance, that technology and management are inseparable and that one cannot study one without studying the other. Perhaps the most surprising insight is his discovery that time studies involved arbitrary, rule-of-thumb decisions that were not scientific at all. "The apparent accuracy and objectivity of stop-watch

time study," he concludes, "was therefore to a large extent
an illusion . . . a ritual whose function it was to validate, by
reference to the apparent objective authority of the clock, a
subjective estimate of the time a job should take" (p. 26).

The subjectivity of this aspect of Taylorism becomes an
important consideration in understanding the molders' reaction to being timed at their work. Indeed, Aitken points out
that their distrust of the Taylor system stemmed largely from
the fact that Dwight Merrick, the time-study expert at
Watertown, knew little about foundry work and made arbitrary decisions about the length of time it should take for a
molder to execute his work. This, coupled with the fact that
Taylor and his military employers remained oblivious to the
feelings and traditions of the Watertown work force, paved
the way for confrontation. Not even the prospect of higher
wages under Taylor's famous premium system could persuade the molders to acquiesce to the new regimen. If anything, it increased their suspicion "that they were being
bribed or fooled into doing something that was not in their
interest" (p. 211). Ultimately Taylorism failed at Watertown
Arsenal because managers neglected to take into account the
customs and feelings of those whom they sought to reform.
Taylor and his disciples had mastered a number of important
technical problems, but they remained blind to the inner
workings of Watertown's social system.

What gives the book its special slant—its originality, if you
will—is Aitken's attention to the arsenal's social and institutional processes. The topic of Taylorism was not new in 1960,
but the way Aitken attacked the subject was. At that time
books written about technological innovation tended to focus
either on the internal development of new technologies or
on their social impact. While Aitken does not ignore these
aspects of the subject, the main thrust of his analysis aims at
understanding the social tensions that arise when new technologies (including management techniques) are introduced

into the workplace. As indicated at the outset, this mode of interpretation would become well known during the 1970s as the "social history of technology."[5] Moreover, his effort to explain the unanticipated consequences of scientific management at Watertown provides an early example of retrospective technology assessment, an area of historical inquiry that is still maturing.[6] Clearly *Scientific Management in Action* is an innovative study, rich in wisdom as well as knowledge. Its evenhandedness in juxtaposing the positions of management and labor make it an exemplary scholarly work. After a quarter of a century it still remains the best study of the day to day aspects of scientific management in action. Princeton University Press is to be congratulated for making this landmark study available again.

Merritt Roe Smith
Massachusetts Institute of Technology

NOTES

1. Frank B. Copley, *Frederick Winslow Taylor* (2 vols., New York, 1923); Sudhir Kakar, *Frederick Taylor: A Study in Personality and Innovation* (Cambridge, Mass., 1970); Daniel Nelson, *Frederick W. Taylor and the Rise of Scientific Management* (Madison, Wis., 1980).

2. Samuel Haber, *Efficiency and Uplift: Scientific Management in the Progressive Era, 1890–1920* (Chicago, 1964). Also see Samuel P. Hays, *Conservation and the Gospel of Efficiency: The Progressive Conservation Movement, 1890–1920* (Cambridge, Mass., 1959).

3. See, for example, Milton J. Nadworny, *Scientific Management and the Unions, 1900–1932* (Cambridge, Mass., 1955); David F. Noble, *American by Design: Science, Technology and the Rise of Corporate Capitalism* (New York, 1977), pp. 257–77; Alfred D. Chandler, Jr., *The Visible Hand: The Managerial Revolution in American Business* (Cambridge, Mass., 1977), Chap. 8, esp. pp. 272–81; Daniel Nelson, *Managers and Workers: Origins of the New Factory System in the United States 1880–1920* (Madison, Wis., 1975), pp. 55–78.

4. See Chandler's review in *The American Historical Review* 60 (1960): 240–41; Gibb's review in *Business History Review* 34 (1960): 273–75; and Rae's review in *Journal of Economic History* 20 (1960): 456–57. Other pertinent reviews appear in *Technology and Culture* 2 (1961): 191–93; *The*

Library Journal 84 (Jan. 15, 1960): 270; and Eugene S. Ferguson, *Bibliography of the History of Technology* (Cambridge, Mass., 1968).

5. See, for example, George H. Daniels et al., "Symposium: The Historiography of American Technology," *Technology and Culture* 11 (1970): 1–35; and Anthony F. C. Wallace, *The Social Context of Innovation* (Princeton, 1982), pp. 3–4.

6. For overviews of retrospective technology assessment, see Howard Segal, "Assessing Retrospective Technology Assessment," *Technology in Society* 4 (1982): 231–46; and Stephen Cutcliffe, "Retrospective Technology Assessment," *STS Newsletter* (Lehigh University), 18 (June 1980): 7–12.

AUTHOR'S PREFACE

A generous endowment of natural resources, a culture that placed a high value on material success and individual initiative, a system of government that left economic power decentralized, a high rate of population growth and capital accumulation, and the ability to borrow and adapt techniques from more advanced countries — these are some of the factors conventionally regarded as having contributed significantly to the economic development of the United States and to the standard of living that we currently enjoy. Typically absent from such listings is any reference to business management. If mentioned at all, it is in connection with the deeds of conspicuous and often notorious businessmen — the robber barons, like Rockefeller, Morgan, or Harriman, or the captains of industry, like Carnegie and Ford. This is like writing the history of warfare in terms of generals and battles. Left out of consideration is the unspectacular work of organization, the day-to-day management of men, machines, and materials. It is strange that in the United States of all places the history of business management and the appraisal of its contribution to economic development should have been paid such scant respect.

The research for this book began while I was a member of the Research Center in Entrepreneurial History at Harvard University. Those of us fortunate enough to work in that group shared a common professional interest in the history of business as a social institution. We respected the work of the business historians, and we did not underestimate the value of the evidence that could be accumulated by the preparation of the histories of individual firms. But we thought that there might be other approaches to the problem. The history of management is the history of ideas, of techniques, of innovations, and of traditions. Development in management is a social process that can be studied in the same way as development in art, in literature, or in science. How have concepts and objectives in business management changed over time? What have been the critical innovations, and how were they accomplished?

To some of my colleagues and to me it appeared that a promising line of attack would be to select a major innovation in business practice, to analyze its content, and to examine its impact upon the ways of doing business that were traditional and normal at the time of its introduction. The Taylor system of management seemed, by these criteria, to be a suitable subject. Here was an innovation that appeared to be clearly identifiable; it could be dated with adequate accuracy; and, to judge by the controversy it aroused, it marked a radical departure from what had gone before.

This book is a study of the Taylor system of management. I have tried to avoid duplicating work that has already been done on the subject. A formal biography of Frederick Taylor is readily available; I have, therefore, felt free to pass lightly over his life and personal activities. Two excellent studies of the reaction of organized labor to the Taylor system have recently been published by Professors Jean T. McKelvey and Milton Nadworny; this topic, accordingly, is dealt with only in outline, to the extent necessary for adequacy of explanation. Nor does this book contain a satisfactory account of the Taylor movement, although that too is a subject of the first importance, essential for an understanding of the origin of the profession of management consultant. What the book does contain is an analysis of the installation of the Taylor system in a particular manufacturing plant, and of the reactions to that installation.

The case method has certain advantages in historical

study: it makes possible an analysis-in-depth that is impossible in a general survey. Regarding Taylorism in particular, case studies are essential for an understanding of the relationship between theory and practice; it is dangerous to base an interpretation solely upon what the Taylor group and their opponents said or wrote about the system. But there are offsetting disadvantages to the case method: one is never quite sure what general significance is to be attached to the particular characteristics of the case chosen. The situation would be much improved if there existed a number of case studies of Taylorism in practice, to make possible systematic comparison. Unfortunately, parallel studies of this type do not yet exist. A few descriptions of plants organized along Taylorist lines are available, thanks to the early work of Hoxie, Babcock, Day, and others; but to the best of my knowledge this is the first empirical account of the introduction of Taylorism — of the process of change as well as the results.

The choice of Watertown Arsenal, a government-owned establishment, may give rise to criticism. The reason for the choice was simply that the source material was immensely richer than for any private firm. From some points of view a private establishment, such as the Bethlehem Iron Company, where Taylor also worked, might have been preferable. The evidence bearing on the Bethlehem case, however, is inadequate for the type of analysis attempted here. At the level of shop management and organization, which is what primarily concerns us here, it may well be doubted whether the fact of government ownership made much difference at Watertown. The institutional reaction to the innovation was admittedly different, because the workers at Watertown had means of protecting their interests not available to workers in private plants. The process of managerial reform, however, and the personal reactions of those involved may not have varied significantly. A workshop in a manufacturing plant has its own structure and problems, no matter where the formal ownership lies.

There remains only the pleasant obligation of expressing my gratitude to those who have aided me in my work. To my former colleagues at the Research Center in Entrepreneurial History I owe a considerable debt — one of which I have become increasingly aware since leaving that hospitable institution. Professor Leland H. Jenks in particular has given me invaluable help and encouragement. For numerous constructive suggestions I am indebted to the friends and associates who were good enough to read the manuscript before publication, and in particular Professor John T. Dunlop of Harvard University and my colleagues at Riverside, Professor Emeritus Gordon S. Watkins and Professor Charles Woodhouse. The librarian of the Taylor Collection at Stevens Institute of Technology provided invaluable assistance, as did the staff of the Army Section, War Records Division, National Archives. Mr. John P. Frey, president emeritus of the metal trades department, A.F.L.-C.I.O., and Mr. Carl Huhndorff, Director of Research, International Association of Machinists, furnished information otherwise unobtainable which must be acknowledged with gratitude. Lastly, and in a more personal sense, I would acknowledge the unfailing encouragement and assistance of my wife, who not only bore the major responsibility for the physical preparation of the manuscript but also, in less obvious ways, made the completion of the work possible. Responsibility for errors of omission or commission, needless to say, must rest solely with the author.

H.G.J.A.

CONTENTS

SCIENTIFIC MANAGEMENT
IN
ACTION

INTRODUCTION

On the morning of February 17, 1915, four men met in an office in Watertown Arsenal, on the left bank of the Charles River, a few miles west of Boston. One of the men was the commanding officer of the arsenal, Colonel C. B. Wheeler, a career Army officer. The other three were civilians: Robert G. Valentine, once instructor in English at Massachusetts Institute of Technology and more recently Commissioner of Indian Affairs and chairman of the Massachusetts Minimum Wage Board, at this time a management consultant in private practice; John P. Frey, vice-president of the International Molders' Union and editor of the *International Molders' Journal*; and Robert F. Hoxie, a brilliant man whose life was shortly to end in suicide, professor of political economy at the University of Chicago.

Probably only Hoxie had much enthusiasm for the assignment that brought them together. Appointed some eighteen months earlier to the staff of the United States Commission on Industrial Relations, he had been assigned the task of investigating the effect on the working man of certain innovations in methods of industrial management, popularly known as Taylorism or "scientific management," which had recently attracted public attention. Some of the work he had already completed: firms using these methods had been identified; lengthy questionnaires had been circulated; and the views of businessmen, management experts, and representatives of organized labor had been conscientiously canvassed. Now, as the final step in his investigation, Hoxie was visiting some of the industrial establishments in which the new methods had been introduced. Watertown Arsenal, for a number of reasons, was high on the list. Valentine and Frey accompanied him in the capacity of "experts," selected by a group of leading management consultants and by the Executive Council of the American Federation of Labor, respectively, to insure the fairness and accuracy of the inquiry. Frey had accepted the job reluctantly, as a public duty he could not shirk. Valentine had been nominated after several of the better-known consultants had declined to serve.

The interview began with Valentine doing most of the questioning.[1] Wheeler was well prepared, with the necessary facts and statistics at his finger tips. He had, after all, been responsible for the managerial reforms at the arsenal ever since they had been first instituted in 1909, and in the intervening years had had to defend them on more than one occasion and before critical audiences. The round numbers in which he described the effect of the innovations were mentioned as if they must surely be familiar already to his listeners. "The increase in output has been one hundred and fifty per cent. We are doing work two and one-half times as fast." Some of the more important changes — the routing of parts and components, incentive payments, time study, and the standardization of equipment — were described and the total effect on production costs estimated as closely as honesty and available information permitted. "The savings in labor cost would be about one-half the labor cost under the day wage system, while the increase in overhead expenses . . . would probably not exceed twenty per cent. . . . We figure that we save in machine work, for every man on premium, approximately $9.00 per day." As an item of incidental interest, Wheeler mentioned that they had recently provided work seats for many of the machinists, for they were confident that under the new system of management there could be no "soldiering" — the popular term at that time for a work pace slower than was thought proper by management.

It was a quiet, factual, unemotional presentation, and the stenographer in the corner of the room might have been pardoned an occasional yawn. Nothing particularly novel or exciting was being said. It was all on the record already — briefly in testimony before the Interstate Commerce Commission,[2] when counsel Louis Brandeis had called witnesses to prove that the Eastern railroads were not being run as efficiently as modern knowledge allowed; at greater length before a congressional committee investigating scientific management;[3] and in infinite detail in the files of the Ordnance Department, in the many progress reports which Wheeler had submitted to his superiors. It was not even particularly controversial any more, for the public press had wearied of "scientific management" and the unions' attacks seemed for the time being to have petered out.[4] Scientific management at Watertown Arsenal was, in February 1915, an established fact. It was now the norm, and an innovation no longer. The days of conflict were, it seemed, past.

As the talk became less formal, Wheeler and the others began to reminisce. For Wheeler and Frey in particular this opportunity for an exchange of information must have been intriguing. A little over three years before they had met under very different circumstances. When Frey had called to see Wheeler in the late summer of 1911, the gates of the arsenal had been patrolled by soldiers with fixed bayonets, and their interview had been much less amicable.[5]

The reason for their meeting on that occasion had been to negotiate a settlement of a strike. The molders employed in Watertown Arsenal foundry had quit work, almost without warning, as a protest against the introduction of time study. As strikes go, it was a very small one, for the men were back at work in a week. But it had become something of a *cause célèbre;* it had precipitated a major political drive against scientific management by the unions; and it had been a serious embarrassment to the Ordnance Department. None of those involved had forgotten it. None of them, even three years later, felt that he understood it.

The topic of the strike arose in the course of the discussion, in connection with the reactions of the employees to the time studies that had been made of their jobs. Wheeler insisted that the workmen at the arsenal had, in the beginning, made no protest against the use of the stop watch. He had had only two individual complaints on that score since the system was instituted, both of them from the same man, a machinist who had been employed at the arsenal for fifteen years, and the matter had been adjusted amicably. The hostility that later arose was due, he implied, to outside influences that had been brought to bear on the men. The fact that the men themselves did not object to time study until union agitation induced them to do so was demonstrated, he argued, by the absence of complaints during the first few months.

Wheeler was repeating what had always been his conviction: that the labor troubles that had afflicted the arsenal since the introduction of scientific management were inspired and instigated by the unions and did not arise from any real dissatisfaction on the part of the men themselves. His statement that he had received only two individual complaints was carefully phrased: the emphasis was on the word "individual." Pressed by Valentine on the point, Wheeler was compelled to qualify his assertion. What he had said was true for the machine shop, where most of the time studies had been made. In that department time studies had been instituted in May 1911, and no complaints had been received from the men until more than six months later, when union agitation against stop-watch time study was becoming extreme. The foundry was a different matter. There trouble had started as soon as the stop watch made its appearance, in the early part of August. The first time study had been made on a certain molder doing a bench job. The man had not complained, and Wheeler and his subordinates had had no idea that there was any objection. The next morning, however, when an attempt was made to time another molder on another job, the man refused. Before Wheeler even learned what was going on, the molders had quit work in a body. Apparently they had held a meeting after work the night before, at which they had decided not to allow any more time studies to be made, and they had prepared a petition on the subject to be handed to Wheeler. But Wheeler failed to see the petition until after the strike had begun. It was a most unfortunate incident, but not really very important. "I think they remained out on strike about ten days, and came back to work again and we continued to make time studies from that time on. . . . We started out just as if they were returning from an ordinary vacation."

Wheeler's plain implication that all the labor troubles which the arsenal had experienced since the introduction of scientific management were due to the irresponsible intervention of the unions was not something that Frey, a union man, could quietly accept. Frey, too, knew something about the molders' strike, and what he knew did not coincide with the impression that Wheeler's remarks had left. Furthermore, there were certain questions about the strike which had never been properly answered, in Frey's opinion. He put a question to Wheeler directly: "I wonder how it is that the men did not get to you before they took any final action in the matter of the strike?"

Here was a weak point in Wheeler's explanation. As commanding officer it was his responsibility to know what was going on in the various departments. Yet this was an instance in which he had clearly been caught unprepared — necessary information had reached him too late. Either of two explanations would fit the facts as Wheeler knew them: the strike might have been spontaneous, an action taken by the men almost on the spur of the moment, without external influence, or it might have been, as he believed, an action taken under instructions from the union, something to be done immediately a certain action was taken by management. Why was he so convinced that the second explanation was the correct one? If it was correct, why had he had no forewarning?

There was only one answer Wheeler could give to Frey's question: "I never found out. As I understand it, they handed the officer in charge the envelope addressed to me, but before it came to me, they went out. I immediately went out to the foundry. When I arrived there, the men had their coats on and were leaving." But was this not inconsistent with his belief that the men were acting under instructions from the union? Did it not suggest rather an unrehearsed, spontaneous action? Not at all, as Wheeler saw it: "To my mind it was premeditated; it was part of a program. They were all pleasant enough that morning. I asked one of the men why he was going out and he said they were ordered to go out. That is all I know about it."

Wheeler's view had, in the years since August 1911, become the official Ordnance Department explanation. It was an explanation which absolved the department of all responsibility. The department had not been maltreating its civilian employees. The fact that a group of skilled men, none of whom had been employed at the arsenal for less than five years, had chosen to quit their jobs rather than work under the new system of management was no reflection on the ability or good intentions of the department. It reflected the irresponsibility and the power of the craft unions.

If plausibility and internal consistency were the only tests, Wheeler's theory might have had much to recommend it. Frey knew, however, that it was wrong, at least in part. The International Molders' Union had no authority whatsoever to order a strike. Its power was limited to giving official sanction to a strike. Furthermore, the giving of official sanction, and the strike benefits that accompanied it, was contingent upon the local union having gone through a prescribed constitutional procedure, involving negotiations with the employer and reports to the Executive Board. This procedure had been completely ignored by the Watertown Arsenal workers. They had gone on strike without even informing their local, far less the International, of what they intended, and the Executive Board of the International had been reduced to giving official sanction to their action after the event. And yet here was Wheeler asserting that the union had ordered the strike. This Frey knew to be incorrect.

To set the record straight, however, it was necessary for Frey to explain how the strike really had occurred, and this he found by no means easy. He wanted to say that the men had gone on strike as a protest against scientific management. But if this were the case, there must have been considerable resentment against the new methods among the arsenal employees. How, then, to explain the fact that the union officials had been as much surprised by the strike as the commanding officer? They had been aware that scientific management was being installed at Watertown; they had publicly berated the new system as being oppressive to the working man; and they had even prophesied that the Ordnance Department would have trouble with its employees if it continued on its course. Had there perhaps been a breakdown of communication between union officials and employees as great as that between employees and plant executives? The explanation Frey wanted to give involved awkward admissions regarding the relations between the unions and their members.

Apart from giving a lengthy exposition of the correct procedure by which a member of the International could officially go on strike, Frey found himself at a loss as to how to correct Wheeler's blunt statements. Valentine put his finger on the crux of the matter: "Apparently no attempt at thrashing this thing out was engaged in before this strike was called?" Frey replied that such was apparently the case. Was this customary? No, it was not customary. It was customary to do everything possible to settle the dispute. In this case, of course, Frey suggested, it being a government plant, and the government having decided to do one thing, and the men having decided that they did not want to do it . . .

But Wheeler would not let this suggestion pass. The men were actually better protected in a government establishment, he pointed out, than they would be in a private firm. They could appeal not only to the officers of the arsenal where they worked, but all the way up to the Chief of Ordnance or the Secretary of War, or they could write to their Congressman or Senator, who could take it up with the President if he wanted to.[6] An employee in a government arsenal was, after all, in a certain sense part owner of the place where he worked. He had means of bringing pressure to bear on management which were not available to an employee of a private establishment.

But if there were all these channels for complaint, why had the men not used them? Well, there was a lot of red tape involved, suggested Frey. Then again, the situation was a novel one. If it had been a matter of a wage reduction, for example, the men would presumably have acted in a more predictable fashion. But time studies were something new and strange, and perhaps they had felt that they had to take a stand immediately or not at all.

Frey and Wheeler were both theorizing. Neither could give a satisfactory explanation of why the strike had occurred. In itself this was of no importance. The strike was long over. Since then the conflict between the Taylor system and the unions had shifted to a different battleground and was being decided by different tactics. Yet it is clear that both Wheeler and Frey were bothered by their inability to produce an explanation that fitted all the facts as they knew them.

It is the unique advantage of the historian that he can sometimes make sense out of past events that did not make much sense to the people who lived through them. This ability, or rather this possibility, has two sources. There is available to the historian more information than was available to the participants, and this more adequate fund of information can be organized in ways which the participants were unable to use. If it were not for these advantages the historian's task would indeed be a futile one. His understanding is inevitably partial, not total; but it need not be as partial as that of individuals who, by reason of their close involvement in the situation, had less information to work with. His explanations of past events cannot but be incomplete; but the passage of time, the accretion of evidence, and the growth of understanding about human behavior make it possible for him to see meaning in what before seemed meaningless.

This book is about the introduction of a particular system of management in a particular manufacturing plant. It deals with the objectives of those who were involved in this innovation, with what they did to accomplish these objectives, and with the degree to which they believed that their objectives had been achieved. It deals with confident and intelligent men who thought that they knew exactly what they were doing and why they were doing it. An effort is made to understand these men, the situation within which they were working, and the point of view with which they approached their task. In this sense the book is an experiment in historical reconstruction. The success of the experiment is to be gauged by the degree to which we gain understanding of the behavior and attitudes of the participants. But the historian can do more than this, if he wishes. He can gain understanding of what these men did not understand, see what they did not see, and find meaning in what they found meaningless. To achieve this he must adopt a point of view different from that of any of the participants, one which will enable him to synthesize all the information available to him and make use of whatever advances in knowledge may have been made between his own time and the time about which he is writing. Thus the historian must have not one point of view but many. He must understand not only what each person thought he was doing, but also what he seemed to the others to be doing. And he must seek some pattern of reasonable meaning in the actions and attitudes of everyone involved.

The introduction of the Taylor system of management at Watertown Arsenal was not merely a technical innovation. It was also a highly complex social change, upsetting established roles and familiar patterns of behavior, establishing new systems of authority and control, and creating new sources of insecurity, anxiety, and resentment. There in microcosm were all the stresses of an industrial society exposed to constant revolution in technology and organization. None of the participants fully understood what they were about; each of them, in different ways, oversimplified the situation. As a consequence, things happened which they could not explain, even to their own satisfaction. Adequate analysis of the impact of Taylorism on Watertown Arsenal requires explanation of the consequences that were unanticipated by the participants and partially inexplicable to them as well as of the consequences that were planned and sought after by the innovators.

Chapter *1*

THE TAYLOR SYSTEM

Questioned by Colonel Wheeler about his explanation of the molders' strike, John Frey admitted that workmen sometimes seemed to behave irrationally. "I know," he said, "the fiendish deviltry with which we throw down our things and go out on strike. They deliberately go in in the morning, and say 'Boys, we'll say "No" to this,' and then they take their time, putting away their things or not, and taking their time about things as they go out." But underlying this apparent irrationality, he insisted, was an attitude that made sense. A walkout was not always a rejection of the job; sometimes it was a means of defending it. "The workman believes when he goes on strike that he is defending his job. . . . Because he has quit, he has simply ceased working, and he is not going to let anyone else take it."

When a man goes on strike, according to Frey, he is defending his job. What did this mean? Implicit in the statement was the conviction that a workman had, in a certain sense, a property right to his job; it was not something that could equitably be changed or taken away by management without the consent, passive or active, of the man involved. Implied, too, was a certain view of what was involved in the concept of "the job." The job was something which could be, or had to be, defended. It involved much more than just a contract for the sale of a certain amount of a man's time, energy, and skill for a certain price. It involved an accepted place, in terms of prestige and power, in the organization of the establishment and of the community; an accepted routine for performing certain accepted tasks; and accepted relation-ships with other individuals and with the physical environ-ment of tools and machines. It involved also a man's feelings: it gave him a point of view from which he could orient him-self toward the world in which he worked and provided him with a stable set of expectations as to how he could count on other people's acting toward him. These elements, and others, were involved in the seemingly simple concept of "the job." All of them were subject to change, and a change in any one of them was a change in the job. Some changes — particularly ones that were unpredictable, that made no sense from the point of view of the individual concerned, or that lowered a man's status in his own eyes and the eyes of others — could be regarded as threats, as dangers to a man's image of himself and his world. Against such threats the job had to be de-fended. One of the possible defenses was withdrawal — the refusal to continue cooperating with the persons from whom the threat emanated.

The idea that the job was something which on occasion might have to be defended against managerial action was not one which would have appealed, or even made sense, to Fred-erick Winslow Taylor. Already in 1915 — the year in which he died — Taylor was generally accepted as the "father" of scien-tific management. Public repute, in this case, was not an in-correct reflection of the facts. Though modestly stating that he personally had contributed little that was truly new and original, Taylor nevertheless regarded himself as an inventor and creator, on the grounds that he was the first to synthesize and systematize the best that was known about the manage-ment of men and to point out the techniques by which this art might be advanced in the future. True, he had not even coined the phrase "scientific management": that had been Louis Brandeis' contribution in the Eastern Rate Case of 1910. Taylor himself had preferred the more accurate but less propagandistic term "task management." But he was the accepted leader of the group of management consultants, engineers, and businessmen who had gathered under the banner of scientific management; his publications on the subject were accepted as the authentic gospel of his sect, at least; the firms which he personally had reorganized according to his principles were regarded as models to be imitated; and in public debate he had taken the position of official spokesman for his cause.[1]

What held the Taylor group together were a vision and a doctrine: a vision of what could be achieved by the application of scientific analysis to the performance of work and a doctrine of approved procedures whereby this vision could be made real. Together, vision and doctrine constituted a cultural innovation of considerable importance — one which has, in little more than half a century, become so much a part of our thinking and activity that today we take it for granted and find it hard to conceive how things could ever have been done differently. Yet Taylor and his colleagues knew that they were innovating. Like all innovators, they tended to overlook their own debt to the past and exaggerate the extent to which they were cutting loose from the pioneers who had preceded them. But in their conviction that they were breaking with tradition, they did not deceive themselves.

Every innovation is a creative synthesis of elements already known. So it was with the concept of scientific management. Essentially it was a late manifestation of the rationalist philosophy, and in this sense it had roots in the eighteenth century and earlier. Prominent too were analogies with Benthamite utilitarianism of the early nineteenth century, in particular the conception of the individual as a social atom, responding to calculations of pleasure and pain that are purely private to himself. What was new in scientific management, or relatively new, was the self-conscious and deliberate extension of rationalism to the analysis of industrial work. It is no coincidence that Taylor and most of his immediate disciples were engineers, for it was by way of engineering that scientific analysis had made its most powerful and continuing impact upon industrial production. They accepted without question the engineering approach that had already proved itself in the design of physical objects, and they extended it to the analysis and control of the activities of people. The essential core of scientific management, regarded as a philosophy, was the idea that human activity could be measured, analyzed, and controlled by techniques analogous to those that had proved successful when applied to physical objects.

So conceived, the idea of scientific management had tremendous implications. Potentially its relevance extended far beyond the organization of industrial work, which was the problem area in which it was first applied. Probably today we have not yet explored the full implications of the concept. Certainly Taylor and his group did not. They were practical men, not philosophers, and their interest was in the immediate problems of industrial production at that time. Thus their conception of scientific management was partial and selective, reflecting the economic and social environment in which they operated. Their preconceptions and their economic interests gave a particular slant to scientific management as they expounded and practiced it. The resistance they encountered stemmed partly from the particular devices and objectives they chose to emphasize; but partly too it was resistance to the basic philosophy of scientific management itself.

Taylor was an aggressive, self-confident individual, seldom at a loss for words and with no small opinion of his own importance. When he expounded the meaning of scientific management, he was capable of making it sound like a turning point in the history of the human race. Yet it is important to distinguish between the idea of scientific management and scientific management as Taylor thought of it and practiced it. In the first place, he was concerned particularly with industry. Some of his colleagues and followers were interested in extending the system to other fields of activity, for instance political and university administration, but Taylor was not. Secondly, within industry he was concerned particularly with metalworking establishments, especially machine shops. In the hands of colleagues and disciples, the system was applied to many other types of business, but it was in machine shops that Taylor did his best and most original work. Thirdly, he was concerned almost exclusively with organization at the shop level — from the superintendent and foreman down. He had nothing to say about finance or pricing or the higher levels of business administration beyond enunciating the so-called "exception principle" — that is, the rule that each level in the administrative hierarchy should concern itself only with matters that are exceptions to the standard procedures of the level immediately below. This is not to say that the principles of scientific management were not thought of as being potentially applicable to a great variety of other problems; on the contrary, Taylor would have insisted that they were universally applicable. The point is that scientific management as Taylor expounded it stemmed from a concern with the organization of work in machine shops. This left an enduring mark on the development of the system.

That Taylor's methods and doctrines were an important innovation can hardly be doubted, but he was far from the first to think seriously about management problems or to try to formulate general principles about the organization of industrial work. In the United States in the period after the Civil War the problem was being attacked from at least three different angles. First, executives and their advisers in large-scale business were concerned with problems of formal organization and control at the administrative level. Leaders in this area were the railroads, notably the Pennsylvania; here we find, for example, discussions of staff and line organization, of executive recruitment and training, and of the formal structure of executive authority. These men were wrestling with the problems of bureaucratic hierarchies and attempting to adapt to the technological imperatives of large businesses the general principles already worked out for other large organizations, such as armies. Secondly, there was the shop management movement. Here the leaders were the mechanical engineers — the men who were responsible for the design and operation of the capital equipment that had come to play such a vital part in American industry. Just as the railroad journals provided the forum for the discussions of the first group, so for the second the publications of the American Society of Mechanical Engineers and other engineering journals were the essential media. Thirdly, there was the reform movement in cost accounting. This was on a slightly different level from the other two, because there is no single group or industry to which we can point as the seedbed of change. The public utilities, operating under the threat or fact of public regulation, naturally showed a special interest in this area; but every man who tried to think or write responsibly about management problems in this period found himself eventually discussing costs and the problems of controlling and measuring costs. Reforms in cost accounting, however, deserve special mention as a source of managerial improvement, for in this area was germinated the seminal idea of costing as a technique not for recording aggregate past performance but for controlling work in process.

Taylor originally made his reputation as a member of the second group: he was part of the shop management movement — an outstanding engineer working along parallel lines with such plant managers as Slater Lewis in Britain and Church, Halsey, and Towne in the United States. He was concerned, as they were, with problems of productivity at the shop level, and he shared their interest in incentive wage payments as a means of increasing productivity. What distinguished Taylor was not that he adopted a radically different approach from his contemporaries and immediate predecessors; rather it was that, accepting many of their assumptions, he carried them to their logical conclusions and embodied the results in an allegedly complete *system* of management that was more inclusive, more self-contained, and more powerful in its practical implications than their proposals and devices. Taylor's system was also much more suitable to serve as the nucleus of a dedicated movement than were the tentative, pragmatic suggestions of other students of the problem. It lent itself excellently, by its drastic oversimplifications and the fact that it could be codified into a relatively small number of rules, to publicity and salesmanship. This is one of the reasons why Taylorism or, as it was later called, scientific management, still evokes recognition and emotional response, while only specialists know the work of men like Halsey and Slater Lewis.

Taylor's interest in managerial reform stemmed from his experiences in industry. While still a young man he was employed as foreman and later as chief engineer at the Midvale Steel Company in Philadelphia, a relatively small engineering and metalworking firm. As foreman his job was to supervise the work of men who were cutting, shaping, or otherwise manipulating metal. Supervision involved seeing to it that certain standards of precision and workmanship were met. At this stage of industrial development this function largely came down to hiring men properly experienced in their craft and firing those who proved incompetent. The foreman's job also involved, however, seeing to it that a certain rate of output was maintained. With given tools and facilities, this implied getting the men to maintain a pace of work that Taylor and his superiors regarded as acceptable.

Taylor quickly became convinced that the men under his charge were soldiering — deliberately restricting output. He took it as his responsibility to break up this practice and increase the pace of work to a rate closer to that of which the physical equipment was capable. He relied at first on the methods normally used by foremen at that time: the threat of discharge, verbal persuasion, and piecework. None of these proved effective. He even recruited some unskilled laborers and taught them the rudiments of the machinist's trade on the understanding that they would back him up in his efforts to increase the work pace. Unfortunately, as soon as they had learned the job, they fell into line with the rest of the machinists and refused to deviate from the norm. Clearly Taylor was being frustrated by the highly effective sanctions operating against any workman who deviated from the normal work pace of his fellows. This frustration, combined with Taylor's compulsion to attain higher labor productivity, led to his initial innovations.

Since this situation was crucial in determining the direction of Taylor's efforts, it deserves analysis. It might be said that Taylor believed that his men were not working hard enough. This, however, would be a misrepresentation of Taylor's views. It was not the expenditure of effort that concerned Taylor. There is no simple one-to-one correlation between expenditure of effort and the rate of work: many other factors influence the relationship — the frequency and duration of waste motions, the number of rest periods, and the degree of nervous tension involved, to mention only the most obvious. Taylor's problem, correctly stated, was to induce his men to increase their pace of work. This was a problem only because of his conviction that their habitual pace of work was inadequate. The crux of the matter was the discrepancy between Taylor's conception of a proper work pace and that of his workmen.

To say that one work pace is proper and another improper is clearly a value judgment. What is possible can on occasion be determined by technological or economic tests; what is proper can be determined only by reference to a given scale of values. When an individual informs us that a certain group of men is working too slowly — that they are slacking or soldiering — he is giving us information about his own values. He is not making a meaningful statement about the group under inspection. The work pace is what it is because the group of persons involved in the work has determined that it shall be so. There is no simple method by which the adequacy, or propriety, or "goodness" of a given work pace can be determined.

Frederick Taylor thought otherwise. He asserted positively that the rate at which work ought to be done could be determined without undue difficulty if the proper procedures were followed, and further, that it could be determined by any person willing to abide by the rules that Taylor stated. This claim was the heart of scientific management as Taylor interpreted it. That such a rate could in fact be determined, and that different individuals observing the same behavior and measuring it by the same techniques would finally determine the same rate as quantitatively correct, was the basic justification for the use of the adjective "scientific."

Taylor's claim was, of course, fallacious. There was, in principle, no difficulty involved in determining the length of time in which a given piece of work *could* be done, assuming that the relevant characteristics of the work, the individual worker, his tools and equipment, and so on were specified. There was even less difficulty involved in determining the length of time in which, on the average, such a job *was in fact done* by a specified workman or group of workmen. But there was no scientific bridge between objective measurements of this kind and the normative judgments that Taylor sought to make. The "can be" and the "ought to be" were not identical. The positive statement was poles apart from the normative judgment.

While at Midvale, Taylor began working out the techniques that were later to become the hallmarks of scientific management. Essentially his method of determining a proper work pace had two elements: job analysis and time study. Job analysis meant that each job was to be divided into particular operations, each of which was taken as a complete unit. The total job was then the sum of these individual operations. In its later developments, particularly at the hands of Lillian and Frank Gilbreth, job analysis was carried to a point at which individual motions of the operator's limbs were taken as the units, and later still the idea began to gain acceptance that the total job was more than the arithmetic sum of its parts. Taylor did not go to these lengths. The primary purpose of job analysis, for him, was to determine the operations that were essential to the performance of the job as distinguished from those that were superfluous or "waste."

This procedure implied a decision, on the part of the individual conducting the analysis, as to what the job really was. Failing this, no isolation of "waste" from "nonwaste" motions was possible. This decision, which to Taylor seemed a simple matter, was in reality a very complex affair. Evaluative criteria entered the picture even at this stage. For example, a machinist assigned the task of turning down a piece of metal to certain dimensions on a lathe might find that his cutting tool required sharpening. To the machinist this might seem a normal part of the job, perhaps a welcome and valued break in routine, certainly a skill that every good machinist prided himself on possessing and exercising. To Taylor, however, or to any of the men he trained, the sharpening of tools was not part of the job. It was *a* job, certainly, but somebody else's job, not the machinist's. To permit the machinist to do it was a waste of his skill and of the money the employer was paying for that skill. Similarly it would not be part of the machinist's job to determine the correct speed of his lathe or the correct angle of cut; even more obviously, it was not part of his job to obtain materials or tools from the

storeroom or to move work in progress from place to place in
the shop or to do anything but turn the piece of metal on his
lathe. Job analysis, as Taylor interpreted it, almost invariably
implied a narrowing down of the functions included in the
job, an extension of division of labor, a trimming off of all
variant, nonrepetitive tasks. This narrowing down was carried out with no consideration for what were traditional parts
of a job and with no appreciation of the possible value to the
operator of actions which seemed superfluous to the observer.
Taylor did not think it necessary to ask the operator what he
conceived his job to include, nor would he have considered
this to be relevant information.

The second basic element in Taylor's system was time
study. After a job had been analyzed into its component operations, these were timed by means of a stop watch. By adding the unit or elementary times for each operation, a total
time for the whole job was calculated. In machine-shop work
and in similar operations it was found useful to distinguish
between machine time and handling time. The former covered the time that elapsed while the piece of work was actually on the machine, with the operative's function confined
to supervision and adjustment. Machine times were susceptible to rather precise calculation, as they depended on
the physical characteristics of the metal being worked, the
cutting instrument, and the machine tool. Frequently machine times could be calculated from data obtained on other
jobs. Handling times, in contrast, referred to the time taken
by the operative in setting the work up on the machine and
removing it after the operation was completed. The sum of
all machine times and handling times gave the total time for
the job.

This total time, however, was not equated directly with
the time in which the job *ought* to be done, which was the
measurement it was desired to find. Instead, certain customary percentages were added to the calculated time to ob-

tain what was called the standard or task time for the job. These "allowances" were added both to the machine time and to the handling time — in the former case to allow for unexpected variations in the hardness of the metal or the speed of the machine, in the latter to allow for fatigue and accidental interruptions. The percentages involved were quite large. At Watertown Arsenal, for example, it was standard practice to add from 25 to 75 per cent to the sum of the elementary handling times and similar percentages to the sum of the machine times to obtain the calculated task time.[2]

Carl G. Barth, one of Taylor's leading disciples, once stated in public: "Our method is called scientific because it determines exactly — scientifically — the length of time in which a man can do a piece of work."[3] In several respects, however, the setting of a standard time for a job involved arbitrary, nonscientific decisions. First, there was the decision as to what constituted the job — what operations were in fact to be timed. Secondly, there was the decision as to which particular men were to be timed, how often, and under what circumstances. When questioned on this point Taylor and his followers usually stated that they would select a "good, fast man," or an "average, steady man," or some similar impressionistic phrase. They would time such a man on the job not once but several times, but the exact number of measurements required to give a satisfactory average was not specified. A man with experience in time study, it was said, would know when he had taken enough measurements. As for the circumstances under which a job should be timed, these were in some respects rigorously specified, but in other respects almost completely ignored. Taylor insisted that all the technical conditions of production should be standardized at the highest attainable level of efficiency before any time studies were made. He saw that there could be no stable measurement of a job time if factors affecting the rate of work were allowed to vary haphazardly. The factors that were relevant, however, he saw as almost entirely mechanical: the speed and feed of the machine, the angle of the cutting tool, the supply of parts and material, and so on. He had no corresponding appreciation of the relevance of social and psychological factors — such as an awareness on the operative's part that he was being timed or a fear that the setting of a task time would result in a cut in pay — nor was any attempt made to reduce the variability of such factors. The furthest Taylor would go in this regard was to recommend that the time-study man should see to it that the operative knew and understood the purpose of the stop watch and that the time study should be made, as far as possible, in a spirit of friendly cooperation.

Lastly, the size of the allowances that were added to the calculated time was decided upon in an arbitrary, rule-of-thumb manner. If technical conditions in the shop had been standardized as Taylor stipulated, there should have been no need to add allowances to the machine time. The only exception rests upon the fact that at the time Taylor was developing his system no convenient workshop method was known for determining the hardness of metals, though laboratory methods were available. The hardness of the metal was thus a variable that could not be completely controlled, and this was a case for making allowances. But with this exception, to add allowances to the calculated machine time was to admit that technical conditions had not been completely standardized and that extra time was being allowed to compensate for an error, whose existence was known but whose magnitude could not be calculated, in the rest of the computation. There was no assurance that this procedure made the total time any less inaccurate than it would have been without the allowances. To introduce one unknown error to compensate for another is hardly scientific procedure.

Allowances were added to the handling time to take account of the fact that, because of varying degrees of fatigue on the part of the operative and uncontrollable interruptions in the flow of work, the time taken even by a "good, steady man" might be greater than the calculated time. To the extent that there were uncontrolled interruptions, this was again an admission that the stipulated conditions for the measurement had not been met. Allowances for fatigue were made for basically the same reason: the operative's accuracy and rate of work varied because of factors not under the control of the man making the time study. Precisely how much time to allow for fatigue was not known, as the problem of industrial fatigue had at that time received no serious study. Here again, the time-study man had to rely on experience.[4]

The apparent accuracy and objectivity of stop-watch time study was therefore to a large extent an illusion. When a task time was set for a certain job, one part of the total had been set by reading measurements from a precise instrument — the stop watch. The other part had been set by a whole series of conventional decisions, in which the values and preconceptions of the individual doing the timing were foremost. It is tempting, though it is only part of the truth, to define time study as a ritual whose function it was to validate, by reference to the apparently objective authority of the clock, a subjective estimate of the time a job should take. Critics of stop-watch time study during Taylor's lifetime often attacked the procedure for its alleged inaccuracies. A more fundamental criticism is that it was never clear exactly what was being measured. Was it the minimum time in which a job could be done? If so, why the allowances, and why not select the fastest man for the measurement? Was it the average time — something Taylor would never have admitted? If so, why the attention paid to the elimination of waste motions and the selection of a particular individual to be timed? Was it, perhaps, an ideal time? In a sense it was. But an ideal — however defined — could not conceivably be measured by a stop watch, nor could it be inferred from any evidence a stop watch could provide. No measuring instrument can produce an "ought to be" from an "is." The so-called inaccuracies and the conventional allowances were not regrettable deviations from scientific precision. On the contrary, they were the means for reaching what was hoped would be a viable compromise between two conflicting work norms: that of the worker and his fellows on the one hand and that of the time-study man, his professional colleagues, and his employers on the other.

Taylor insisted that, for stop-watch time study to be carried out correctly, all the conditions of work should be standardized at the highest possible level of efficiency. His conception of what conditions of work were relevant and therefore demanded standardization was a limited one. Nevertheless, his search for methods of standardization produced some of his most solid and remarkable achievements. For example, at the time Taylor was doing his pioneer work the machines in most factories were run by systems of belts, pulleys, and shafts from a central power source; the individual electric motor was not then in general use. The speed of each machine, and therefore the work pace of the operator, depended directly upon the mechanical efficiency of the belt and pulley system. Incorrect tension on the belts caused slippage; inadequate maintenance led to breakdowns and sometimes to serious injury. What was required, as Taylor saw it, was first a method for determining scientifically the correct tension to be put on a belt and the correct methods of maintenance, and second the establishment of belt maintenance and adjustment as a separate job, to be attended to by men trained in the work under special supervision.[5] Here was, in microcosm, Taylor's basic method in action: the analysis of a piece of work, the determination of the best methods of handling it, and the reorganization of the division of labor in the factory so as to insure that these methods were applied. Typically, it involved

the divorce of ancillary functions from established jobs —
ordinarily no one in particular looked after the belting — and
the establishment of new jobs.

To the layman there was nothing very remarkable about
this: surely an idea like this could occur to anyone? Indeed it
might, and probably some such arrangement had occurred to
other shop engineers before Taylor. But to Taylor it was not
an incidental detail; it was something that *had* to be done
if time study was to be feasible. Like all factors influencing
the pace at which work was done, it had to be studied and
standardized. His studies of belt maintenance and adjustment were an integral part of his attempt to secure total control over the pace of work.

Precisely the same analysis can be made of Taylor's other
contributions to shop management; all were designed to
achieve total control of the job and its performance and in
particular to enable management to prescribe and enforce
a standard work pace. With the advent of the individual elec-
tric-motor drive — revolutionizing as it did the whole art of
factory layout — Taylor's work on belting lost much of its
practical significance, though it remains a classic example of
the controlled experiment applied to shop engineering. Other
problems to which he turned his attention, however, produced solutions of more enduring significance: the idea of
planned routing and scheduling of work in progress, fore-
shadowing the techniques of the assembly line and con-
tinuous-flow production, with the route of each assembly and
subassembly scheduled in advance from a central planning
room; the introduction of systematic inspection procedures
between each operation; the use of printed job and instruction cards, informing the worker in detail what operations
were to be performed on each component and informing
management precisely how much of each machine's and
man's working time was devoted to each product or batch of
products; the introduction of refined cost-accounting tech-

niques, based on the routinized collection of elapsed times for each worker and machine on each job; the systematization of stores procedures, purchasing, and inventory control; and the concept of "functional foremanship," with its four or five supervisors with specialized functions in place of the normal multifunctional single foreman.[6] Of all these innovations, only functional foremanship failed to win general acceptance. The remainder quickly became standard workshop practice and, with modifications, have remained so to this day. Because there was nothing spectacular about any one of them, and because they were all organizational changes which could not be patented or branded with the Taylor name, there has been a tendency on the part of historians of industry to play down their importance. In this respect history has done Frederick Taylor less than justice, for these inconspicuous innovations have probably exercised a more far-reaching influence on industrial practice than has the conspicuous innovation of stop-watch time study.

There is one of Taylor's innovations that deserves special comment, not only because of its relevance to what happened at Watertown Arsenal but also because it furnished Taylor with considerable prestige among industrialists and engineers for reasons independent of his strictly managerial reforms. This was the invention, or rather the discovery, of high-speed tool steel. The primary object of Taylor's endeavors was to find a means of determining the rate at which work should be done. Every factor which could influence the pace of work had to be brought under control and standardized at the optimum level of efficiency. In the machine shops where Taylor did most of his early work, a factor of obviously crucial importance was the speed at which the machine tools were run. Taylor was therefore confronted with the problem of determining the proper speed at which any given metal-cutting tool should be operated. This was a problem which, up to that time, no one had tackled systematically.

What Taylor was after was not a new tool steel; he was concerned rather to find out which of the tool steels then available was the best, in order that he might take it as his standard. The crucial series of experiments were run by Taylor and his colleague Maunsel White, a metallurgist, at the Bethlehem Iron Company in 1898. Experimenting with different methods of heat treatment, Taylor and White discovered that cutting tools made of steel containing 7.7 per cent tungsten and 1.8 per cent chromium (high-speed steel, as it was called) attained their optimum cutting efficiency at temperatures just below the melting point of the steel. A cutting tool made of high-speed steel operated at maximum efficiency when run at the highest speed possible without melting the steel.[7]

What had Taylor and White discovered? A new tool steel? Certainly not: the chromium-tungsten steels with which they were working had been purchased commercially from the Midvale Steel Company. A new method of heat treatment? Possibly, for up to that time it had been generally accepted that a tool heated to a temperature above cherry red was permanently ruined. But much more than this was involved. What Taylor and White had done was to show how the new alloy steels could be used to cut metal at rates several hundred per cent faster than had been possible before; they had opened the way to a revolution in machine-shop practice. Incidentally, and this was probably Taylor's contribution rather than White's, they had upset one of the most hallowed precepts of the machinist's craft: the belief that the proper cutting speed was the one that maximized the life of the cutting edge of the tool. Not so, said Taylor: tools can and should be reground regularly and systematically; what we should maximize is the amount of metal removed per unit of time, and cutting speeds should be set accordingly.

The Taylor system of management in later years came to be known by the labor unions as a "speed-up" system. The phrase had strong emotional connotations and did less than justice to what Taylor was trying to do. Taken as referring solely to what Taylor called "the art of cutting metals," however, it is literally correct. Taylor's experiments convinced him that most machine tools used in American industry were being operated too slowly. If a machine tool was being operated at its correct speed, it was purely by chance; not one machinist in a thousand, said Taylor, realized that there existed clearly defined laws governing the speed and duration of cut.[8] Cutting speeds that might have been not too inappropriate at an earlier day, when straight carbon steel was the only tool steel available and when the horsepower applied to each machine was smaller, were still being used in machine shops that had available vastly greater horsepower and improved tool steels. The reason, as Taylor saw it, lay in ignorance of the scientific laws of metal cutting — an ignorance characteristic not only of wage-earning machinists but also of machine builders and mechanical engineers.

Taylor once expressed the relationship between his experiments on tool steels and his system of management as a whole in these terms: "The moment that scientific management was introduced in a machine shop," he said, "that moment it became certain that the art or science of cutting metals was sure to come." [9] This statement certainly expressed one aspect of the relationship. If Taylor was to attain his primary goal of securing complete control over the pace of work, it was essential that he know precisely the maximum rate of output of which his machines were capable. This in turn involved careful analysis of elements in the work situation which previously had been left to tradition and common sense.

There is, however, another way of looking at the relationship between the discovery of high-speed steel and the development of scientific management. This was expressed concisely by James Mapes Dodge, president of the Link-Belt

Engineering Company, a warm friend of Taylor's and a close observer of the development of Taylor's ideas. "The whole development of Mr. Taylor's work," wrote Dodge in 1909, "is based on the ability of high-speed steel to remove more metal per minute than any previously used or discovered steel could do." [10] The implication of this statement is that scientific management was essentially a method for exploiting the full productive potentialities of the new cutting tools. The discovery of high-speed steel marked a sharp discontinuity in machine-shop practice. Machines could now be run at from two to four times their former speeds. The organization of the machine shop, and indeed of the whole factory, had to be adjusted to this new level of productivity.

Historically, Dodge's statement is incorrect. Taylor had begun his work with job analysis, time study, and incentive payments before his discovery of high-speed steel. The experiments with tool steels were carried out because time study required the standardization of machine-tool practice. Nevertheless, Dodge's words show considerable insight. Without high-speed steel Taylor's managerial reforms might have been highly desirable; with high-speed steel they became well-nigh indispensable, at least in machine shops. The innovation of high-speed steel spread much more quickly through American industry than did the innovation of scientific management. Its advantages were more obvious, its nature more familiar, and its adoption more easy. Employers and wage-earning machinists alike might well be suspicious of Taylor's managerial reforms, but they would find it much more difficult to resist the introduction of a superior piece of technology. But if high-speed steel was to be effectively utilized, scientific management or something very close to it had to be adopted too.

The discovery of high-speed steel did not solve the problem that Taylor had originally set himself. Indeed, it raised new problems. By reducing machine times by a half or more,

it focused attention on the proportionately greater importance of handling times. Efforts to reduce handling time by
job analysis and time study therefore were increased. And it
now became even more urgent than before to find a method
for determining exactly, for any given piece of work, the correct speed and feed of the machine tool.

In tackling this latter problem Taylor had at Bethlehem
the help of two men who later became leading exponents and
practitioners of his system: Henry L. Gantt and Carl G. Barth.
Gantt had earlier worked with Taylor at Midvale and was
familiar with what he was trying to do. With his assistance,
Taylor succeeded in determining empirically, by a prolonged
series of experiments, the optimum relationship between all
the variables that influenced the rate at which metal could
be cut on a lathe: the depth of cut, feed, speed, and type of
tool, the hardness of the metal, the power applied to the
machine, and so on.[11] These results were plotted on graph
paper, giving a set of geometric curves from which the proper
speed of the lathe could be determined when the values of
all the other variables were known. This method of solving
the problem was, however, too slow and inconvenient for
ordinary workshop use. Barth, soon after he came to join
Taylor at Bethlehem in June 1899, reduced the relationships
discovered by Taylor and Gantt to a mathematical equation
and transferred the functional relationships involved to specially made slide rules, which made it possible to determine
the correct speed of a machine tool quickly and with all the
accuracy required for practical use. Slide rules of this type,
usually made by Barth personally, became in time a standard feature of the Taylor system. Like that other hallmark
of the Taylor system, the stop watch, Barth slide rules provided management with a quantitatively precise and ostensibly nondebatable method for determining the rate at which
work should be done. Whereas in stop-watch time study,
however, an element of human motivation was involved —

the operator had somehow to be induced to work at the pace prescribed — all the variables included on a Barth slide rule were subject to mechanical control.

The industrial worker, suggests Daniel Bell,[12] lives within a set of constraints imposed on him by three related logics: the logic of size, dictating the concentrating of men at a common place of work; the logic of metric time, which sets his task and determines his pay by reference to the clock; and the logic of hierarchy, which concentrates in management all significant decisions as to the job and how it shall be performed. These three logics are made inescapable by the complex division of labor characteristic of modern industry; they determine the nature of work in the plant and indirectly the nature of life in industrial society. Scientific management, as expounded by Taylor, was in effect an attempt to translate into a set of practical managerial procedures these logics of size, time, and hierarchy.

Taylor's insight into the general logic of industrial work, however, was combined with particular assumptions about human psychology and a particular approach to the problem of incentives. Taylor, like almost all his contemporaries, accepted with little question the view of human nature implied by Benthamite utilitarianism. This acceptance determined Taylor's approach to the problem of incentives in the workshop. It was necessary to create an environment within which the individual, by rational calculation of what was to his own advantage, would do that which management wished him to do. Such an environment, in its mechanical and administrative aspects, could be created by the many detailed reforms in shop practice that made up the Taylor system. There remained the problem of providing a direct and powerful incentive to the individual worker to induce him to put forth his best efforts — to create that spirit of friendly cooperation which Taylor insisted was essential to the success of his methods. The goals of the worker had to be brought into coincidence with the goals of the enterprise. This problem was to be solved by gearing the worker's wage to his production through a system of incentive wages.

Taylor had his own incentive-payments system which he had devised and which he regarded as more powerful than any of the alternatives: the differential piece rate. He realized that no incentive-payments system could by itself bring about important increases in productivity in the absence of the other reforms in shop management that he advocated; nevertheless in his early expositions of his system — specifically in papers presented to the American Society of Mechanical Engineers — the differential piece rate was made the central feature of the presentation, other reforms in shop management being introduced as prior changes that were essential if the differential piece rate were to function effectively. This emphasis on incentive wages was to have important consequences in molding the public image of the Taylor system; but, as an engineer speaking to and writing for other engineers, Taylor had to phrase his presentation in terms that would be familiar and acceptable to his audience. In this case he knew what he was about.

Founded in 1880, the American Society of Mechanical Engineers during the first six years of its existence concerned itself with problems relevant to the duties of mechanical engineers as these had been traditionally conceived — that is, with questions of machine design and operation. The span of discussion did not begin to enlarge until 1886, when for the first time papers on problems that were clearly managerial in nature were presented. Chief among these was a short but brilliant paper by Henry R. Towne, president of the Yale and Towne Manufacturing Company. Bearing the significant title, "The Engineer as Economist," [13] the paper stressed that the engineer employed in industry could no longer with impunity concern himself solely with mechanical efficiency —

that is, with doing things in the best way by engineering standards. He had to recognize other criteria of efficiency, and in particular of economic efficiency, expressed in terms of costs and revenues. If engineers employed in industry, argued Towne, are to be called upon to function as executives rather than, or in addition to, their functions as engineers, then they must widen their intellectual horizons and learn to think and act as economists, or in other words be aware that business decisions necessarily involve consideration of money costs.

But how was an engineer to learn to be a manager? Where was the body of accepted theory and practice to which he could refer? Could management be learned and taught and reduced to standard principles as engineering was? No, Towne admitted, it could not — not, that is, if the art or science of management was permitted to remain in its then inchoate and rule-of-thumb condition. What was needed was analysis and systematization: a search for the best practices, an analysis of what worked in management, a beginning with the task of deducing general principles of management analogous to those of sound engineering. A vast amount of accumulated experience in shop management already existed, but it had never been systematically collected and generalized. Each established business went along in its own way, receiving little benefit from the parallel experience of similar enterprises; while each new business gradually developed its own methods of operation, learning little if anything from all that had been done previously. To serve as a clearinghouse for the best available information on managerial practice would be, Towne suggested, a fitting function for the A.S.M.E.

Towne's analysis was penetrating, but his recommendations were general. From the starting point he had provided, progress could have been made in a variety of directions. Actually the men who followed Towne's lead at later meetings of the A.S.M.E. tended to concentrate on one specific problem: the design of incentive-payments schemes. There was nothing in what Towne had said to direct their attention to this managerial device rather than others, beyond the general injunction to pay heed to the cost calculus. The explanation must be sought in the practical problems the members of the A.S.M.E. were encountering in their industrial work and the way in which they interpreted these problems. The last quarter of the nineteenth century was a period when heavy capital equipment was coming into use on an unprecedented scale in American industry. The substitution of capital for labor in production brought tremendous increases in productivity — and a new importance to mechanical engineers. Increases in productivity, however, were limited by the carry-over from an earlier period of crude methods of work training and organization at the shop level. American industry was relying for its skilled labor upon traditional craft training and for its unskilled upon mass immigration. Shop management, now that factories had grown too large to be supervised by a single owner or his agent, was left almost wholly to foremen, recruited from the ranks of wage labor and unassisted by specialized staff except in strictly engineering matters. However sophisticated American industry was becoming in matters of finance and corporate strategy, and however complex in technology, at the level where in the last analysis the work was done and the goods produced, its organization and methods remained primitive. The uneven quality of the work force, reliance upon immigration for new recruits, the relative shortage of supervision, and the growing complexity of division of labor in the workship — these factors, among others, encouraged an intense interest in incentive-payments systems. Consistent as they were with the American business ethos of competition and the profit motive, piecework and its variants appealed strongly to industrial executives; to industrial engineers such systems seemed to offer the dynamic they needed to translate machine potential into man-hour production.

Concerned professionally to design and operate machines that would run efficiently, and reminded by Towne that efficiency was a matter of dollar costs as well as of minimum energy loss, the engineers of the A.S.M.E. — or those of them whose interest was aroused — interpreted the problem facing them in the same terms they would have applied to a problem in technology. The machine — the lathe, the loom, the rolling mill, or the drill press — could produce so many units of output in twenty-four hours. Ordinarily it produced less than this, and unit costs were higher as a result. The way to reduce costs was to operate the machine at capacity, or as close to it as possible. Some concessions had to be made to human frailty and social custom. It was not ordinarily possible to run a machine twenty-four hours a day, though as far as the machine alone was concerned, assuming repairs and maintenance were adequately provided for, this would represent optimum economy. More realistically, it was not usually possible to operate the machine continuously at its maximum efficiency even during working hours. The reason for this was not a technological one; it involved, in the words commonly used at the time, "the human aspect." Workers apparently did not wish, and could not be induced by any normal methods, to work at the pace of which the machine was capable.

This was a psychological problem, not a technological one. The mechanical engineers who tackled it fell back on the cruder versions of the crude psychological knowledge of their day. The worker, like every other human being, was motivated by rational self-interest. Self-interest, in this case, meant the monetary reward he received for his work. To get him to work harder, all that was necessary was to offer him more pay on condition that he did so. This resolved itself into the problem of designing incentive-payments schemes — systems of buying labor that would make the total daily pay received by the worker depend on his daily output. This was a problem by no means uncongenial to men with engineering training and experience: it was quantitative and mathematical; it was susceptible to refinement in detail; and, by its apparent practical adequacy, it made superfluous any attempt to explore more comprehensively the factors that influence motivation. A problem of considerable psychological complexity was to be disposed of by a mechanical linkage of pay to productivity.[14]

To anyone looking at the problem of incentives from this point of view, ordinary daywork had little to recommend it, since it gave the worker no financial interest in increasing output. Straight piecework was apparently ideal, since it related pay received to volume of output in a direct one-to-one fashion. In practice, however, it had proved disappointing. Employers had consistently found it necessary to cut the rate of pay per piece as their wage costs rose, and employees, expecting such cuts, had acquired the habit of holding back so as not to "spoil the job." The incentive effects were lost, because as soon as a pieceworker began earning more than his employer expected, the rate was cut. Efforts were therefore directed toward designing more complex payment systems, in particular ones that would obviate the necessity for rate cutting. A system of this type had to have two characteristics: the total wage received by the worker had to increase in proportion to increases in output, though not necessarily in direct proportion, and labor costs per unit of product had to fall, or at least not increase, as output rose.[15]

Henry Towne presented such a plan to the A.S.M.E. in 1889.[16] Christened "gain sharing," it involved essentially the drawing up of a contract between employer and employees, to last at least one year. This contract was to be based upon a division of costs of production into those that the worker could influence by his own efforts and those that he could not. Any savings achieved in the former class of costs would be divided up at the end of the year between the management (50 per cent), the foremen (10 to 15 per cent), and the workers (40 to 35 per cent). Towne made two stipulations, however: first, costs should be reckoned, when the contract was first drawn up, not at the true book costs to the company but at a figure from 10 to 30 per cent less than this, so as to enable management to hedge against the possibility of substantial subsequent reductions; and secondly, when one contract expired, management should have the right to cut the rates (*i.e.*, the "costs") in the subsequent contract.[17]

More direct than Towne's scheme was an incentive-payments plan presented to the A.S.M.E. by Frederick A. Halsey in 1891, a significant feature of which was that it was based on job times.[18] The time required for any given job was known from previous experience. For any savings made in job times, the worker was to be paid, in addition to his daily wage, a new premium rate at so much an hour for the time saved. This premium rate was always less than the daily wage rate. One third of the day rate was usual, so that for each hour saved, one third of the gain accrued to the employee. Any increase in the rate of production would certainly increase the worker's total pay; but it would also reduce labor costs per unit, since for a marginal increment to output the worker would be paid at only a fraction of his daily rate. This, Halsey argued, would obviate the need for careful cost keeping (required by Towne's scheme), for the setting of individual rates per piece (a delicate and often acrimonious procedure), and for cutting rates when output rose (since average labor costs per unit would inevitably decline). The historical level of output was taken as the basis for the whole calculation.

Frederick Taylor presented his first formal paper to the A.S.M.E. in 1895.[19] At this time he had been working on the development of his system of management for more than ten years, but it was only by indirection that his work on time study, routing, standardization, and so on appeared in the paper. Entitled "A Piece-Rate System, Being a Step Toward Partial Solution of the Labor Problem," it presented an incentive-payments scheme that appeared on the surface to be essentially an elaboration of Halsey's premium plan, since it was based on job times. There were, however, certain significant differences. Job times were to be determined not from past experience but by stop-watch time study. Standard times were to be set for each job and a standard rate of output thereby determined. This was to mark a sharp discontinuity in the rate of pay. Workers who took a longer time to complete the job were to be paid at a very low piece rate — a rate so low, in fact, that it would be hardly possible to earn a regular day's pay. Workers who completed the job in a shorter time would be paid at a higher rate per piece. Needless to say, if the time study had been carried out correctly, it would be almost impossible for a worker to do better than the allotted task time, so that there was a built-in guarantee that no man could ever earn excessive wages.

Taylor argued that, under his plan, there would never be any need to cut a piece rate. Rate cutting, he said, had been necessary in the past only because employers had had no means of knowing how fast a job ought to be done. The time required for a new job was always overestimated; then, as the workers became more familiar with the job, they began to earn excessive wages and the original rate had to be cut. Determine the time required correctly in the first place, said Taylor, and you will never find it necessary to change the rate, because you can be sure from the start that the rate of work you set as the norm will never be exceeded.

In some respects Taylor's claim appears rather ingenuous. There were at least two ways in which, under his system, management could reduce piece rates. First, Taylor's plan stipulated only that a higher rate should be paid for work done in less than the task time and a lower rate for work done in more. It laid down no rules for what the higher rate should be. Taylor's presentation, however, made it clear that this so-called higher rate was in fact to be smaller than the regular piece rate.[20] Even to maintain his daily wage at the previous level, far less increase it, the worker was compelled to increase his output substantially. On this basis the very installation of Taylor's differential piece rate might seem, from the worker's point of view, equivalent to a cut in the rates. Secondly, management invariably retained the right to change the rate per piece if any changes were made in the way the work was done. If, for example, a worker operating a drill press was provided with a jig to guide his drill, the rate per piece would be reduced; similarly, in machining work, if any changes were made in the speed of the lathe. In general, any reduction in machine times brought with it a reduction in the piece rate. Some such provision was necessary if management were to have any incentive to adopt new methods. But possibilities of abuse lurked in the background. What was to stop a cost-conscious foreman from making some slight change in the job purely in order to cut the rate? What was rate cutting from one point of view was not necessarily rate cutting from another.

It is clear that the mere introduction of a differential factor in the piece rate and the mere setting of a standard work pace by time study did not in themselves determine what the piece rate should be. This was a matter on which Taylor was capable of deceiving himself, for on more than one occasion he publicly stated that his system of management made possible a scientific determination of how much a worker should be paid.[21] This it certainly did not. Taylor's differential piece rate did involve assumptions as to how large an increment in pay was required to call forth a given increment in output. But it shed no light on how the basic rate was to be determined.[22]

Taylor regarded his differential piece rate as the most powerful type of incentive-payments system that could be devised. It gave the worker a strong and immediate incentive to increase his pace of work; it protected the employer from undesirable increases in average labor costs; and, by penalizing men who failed to maintain the standard output, it gradually weeded out from the work force all but "first-class men." These were weighty advantages for any employer. Nevertheless, the differential piece rate was never widely adopted. Some of Taylor's colleagues, such as Gantt, preferred other systems of their own devising, less powerful and punitive than Taylor's; and even Taylor on occasion approved the use of such earlier and milder methods as the Halsey plan. The differential piece rate, to function effectively, required a degree of standardization and systematizing that could not always be achieved, particularly in shops where the product or the methods of production changed frequently. Then too, it was more complex to administer than straight piecework or the Halsey plan, and the advantages to be gained from the added complexity were not always clear. But in addition, the differential piece rate violated the worker's sense of equity and was almost without exception vigorously opposed by the unions.

It is not easy to generalize about the attitude of the unions to incentive-payments schemes in this period. The consensus of those who have examined the problem is that a distinction can be made between straight piecework and more complex systems such as Taylor's. Opposition to piecework was far from universal: it was preferred or willingly accepted by the coal miners, the textile workers, the cigar makers, and a number of others. The molders' union struggled to drive piecework out of the jobbing foundries, but accepted it in the stove and furnace industry. The International Association of Machinists officially prohibited the acceptance of piecework, but many of its members preferred it and negotiated piecework agreements through their own shop representatives. In general, piecework was accepted and indeed preferred where the unit of output could be defined with precision and where conditions of work could be maintained with substantial uniformity over periods of time. Where both these conditions were met, the workers were likely to prefer piecework to straight time work: their earnings varied directly with their effort, and they were protected from the danger of a speed up without additional compensation, which was possible under the day wage system. Where either or both of these conditions were not met, piecework was likely to be opposed both by the workers and by their unions.

Opposition to the differential piece rate and the premium and bonus systems was both more widespread and more determined. In contrast to straight piecework, which rewarded an increase in output with a proportionally equal increase in pay, these variants provided for a reduction in the rate per piece once the standard output was exceeded. As Slichter puts it, "The fast workers [were] compelled automatically to cut their own piece rates."[23] The concept of "a fair day's pay for a fair day's work" ruled out any incentive system under which a given percentage increase in output over a standard amount was not rewarded by at least an equal percentage increase in pay. The base rate was taken as the standard: a "thirty-cent job" should remain a thirty-cent job, no matter what the worker's rate of output might be. Thus the very feature of such plans that commended them to the employer made them unacceptable to the unions, since they ran counter to the whole philosophy of a standard rate for each job. Unions which accepted piecework rejected premium or bonus plans, while unions which found even piecework unacceptable regarded premium or bonus systems as anathema.[24]

The unions' opposition to his particular type of piece rate, as to his other innovations, helped to confirm Taylor in his conviction that he was right. Unions he regarded as essentially pernicious institutions, for they were responsible for the systematic restriction of output which to him was the unforgivable sin.[25] Scientific management, he was convinced, would do far more for the workers — if only they would "cooperate" — than unions ever had done or could do, for scientific management was based on an understanding of the laws of production, not upon the opinion of the ignorant. These laws could not be bargained over. In particular, there was, Taylor believed, no justification whatever for the doctrine of the standard wage. Insistence upon the payment of the same wage to all workers of a given class, regardless of their contribution to the productivity of the plant, meant that the most efficient and energetic were penalized for the benefit of the inefficient and lazy. This was not only undesirable from the point of view of the employer, it was also ethically wrong.

In contrast to the standard wage doctrine, Taylor claimed that his system of management treated the workers as individuals. It was therefore more democratic. If the Taylor system was adopted, the individual worker was provided with all the instruction he could require. His proper job was explained to him; he was shown the best way to do it; he was furnished with tools properly maintained and machines properly adjusted; and he was paid in proportion to his output. What could be fairer than this? What could give a worker more self-respect than the knowledge that, while everything that could be done to help him had been done, beyond this he was on his own? If he produced more, he would be paid more. If he failed to produce more, but was able and eager to learn, he would be shown the way. All that was necessary was for him to approach his job in the right frame of mind and to be willing to cooperate with those who knew how to help him.

Cooperation meant different things to Taylor in different contexts. For the workers (in Taylor's own words) it meant "to do what they are told to do promptly and without asking questions or making suggestions." [26] Whatever changes Taylor or his colleagues might make in the job, the worker was to accept them without question. His role was to be passive — that of an efficient but self-effacing servant of the real producer, the machine. Human ability and motives, not the potentiality of the machine, set the ceiling on the feasible work pace. To raise this ceiling, Taylor relied on the inducement of the dollar. A carefully engineered incentive-payments scheme, operating within a work environment routinized and standardized in every detail, was to overcome the final barrier to increased productivity: the traditional work pace of the worker and his mates.

General William Crozier, Chief of Ordnance, carefully feeling his way toward the adoption of the Taylor system in the government arsenals, once asked Taylor, with disconcerting directness, why he laid such stress on incentive payments.[27] Had not Taylor assured him repeatedly that the men would have to work no harder under the new system, that they would in fact find their work much easier and more pleasant? Why then pay them more? Taylor replied that it was true that a great deal of time could be saved by such things as eliminating waste motions, insisting on the proper sequence of movements, and so on; but that, even after all this preliminary work had been done, it was still necessary to induce the men to exert their skill and energy to full effectiveness, so that each one of them could attain "the quickest time in which a job should be done by a first-class man." As for paying them more, Crozier should have no worry on that account. A worker might on occasion do more than was asked of him (*i.e.*, exceed the norm), but it would be most unusual. "It may happen once in a thousand times . . ."

It was not because Taylor felt that men deserved higher pay for more production that he stressed incentive-payments schemes. Such a system, and preferably one that punished low output as well as rewarded high, was the necessary keystone to the Taylor system of management. Whatever changes were made in the institutional and mechanical framework within which the job was performed — reorganization of stores and supply procedures, planned routing of work, systematic inspection, instruction, maintenance, and even the respeeding of machines — none attacked directly what Taylor had from the start believed to be the principal cause of low output: soldiering. Only an incentive-payments scheme, as Taylor saw the matter, could do that. After all the preparatory work had been done, then was the time for the carefully planned break-through that would shatter the deeply entrenched work habits of the wage earner.

What Taylor called soldiering, however, was by no means as simple a matter as he imagined. Ignorance, malice, stupidity, false doctrine spread by the unions, memories of rate cutting:[28] these were, in his mind, the only possible explanations of why workers deliberately produced less than they knew they could. Nothing in his upbringing, education, or experience, nothing even in the practical or academic knowledge of his time, furnished him with the concepts that might have made possible a different view of the problem. But restriction of output is not restriction unless measured against some relevant standard, and the standards of the engineer and employer have no more claim to absolute validity than the standards of those who are alleged to be doing the restricting. Rates of work are not determined by chance or by the quirks of individual personality.[29] The idea of a norm of output as a universal feature of all organized groups, set and maintained by the group itself and often defended by highly effective sanctions, would, however, have been entirely alien to Taylor, as to most others of his time. Highly sophisticated in its approach to problems of formal organization and technology, in its analysis of motivation and group behavior the Taylor system was crude in the extreme.

Chapter 2

THE ORDNANCE DEPARTMENT

Innovations in technology and managerial methods, no matter how far-reaching their effects in the long run, are seldom so dramatic as to make an immediate impress on the mind of the public. Often they are highly technical and can be fully understood only by specialists. Often, too, unless connected with techniques of mass destruction, they lack the element of human interest that the newspaper requires. So it is that the new ideas and devices that shape the development of a culture frequently remain almost subterranean, as far as the attention of the bulk of the public is concerned, and produce their effects quietly.

The Taylor system of management, in its early years, was such an innovation. Taylor himself, until he acquired a taste for fame, showed little interest in broadcasting his ideas to the world. And at first the world knew nothing of him. Two events combined to make the Taylor system a matter of public concern. The first was unexpected: Louis Brandeis' introduction of "scientific management" in the Eastern Rate Case of 1910, which for the first time brought Taylor and his associates to the witness stand and made their testimony news. The second was the culmination of many years of planning and negotiation by Taylor and his friends: the adoption of the Taylor system by the Ordnance Department of the United States Army for use in the manufacturing arsenals, in 1909.

The decision to adopt the Taylor system of management was taken by General William Crozier, Chief of Ordnance.

Crozier, who was graduated from the United States Military Academy in 1876, was a man of many talents and wide experience. He had seen active service against the Sioux and Bannock Indians, in the Spanish-American War, in the Philippines campaign of 1900, and in the relief expedition to Peking in the same year. He was a delegate to the International Peace Conference at The Hague in 1899; during World War I, he was a member of the Supreme War Council. In the quieter moments of his life he had been an assistant professor of mathematics at West Point, and he had acquired a certain reputation as an inventor, being responsible not only for a wire-wound gun of novel construction, but also, with the aid of General Buffington, for an ingenious disappearing gun carriage, intended principally to help defend the seaports of the United States against the invasion of foreign powers.[1]

From 1901 until 1917, with the exception of one year (1912–1913) when he was president of the Army War College, Crozier held the rank of Chief of Ordnance of the United States Army. In this capacity he was responsible not only for the designing, testing, and storing of all ordnance equipment and for the allocation of ordnance among the various units of the Army, but also for the management of five government-owned manufacturing arsenals: Rock Island, Frankford, Springfield, Watervliet, and Watertown. The executives of these arsenals were officers of the Ordnance Department, and a few enlisted men were attached to each for guard duties. The work force, however, was made up of civilian employees, hired through civil service procedure. The arsenals manufactured a wide range of products, ranging from small-arms ammunition at Frankford to seacoast gun carriages at Watertown. Funds for their operation came from congressional appropriations, principally under the Army Act and the Fortifications Act. They were not, of course, the only suppliers of ordnance material to the

United States Army, but were normally in competition with private concerns such as the Bethlehem and Midvale Steel companies. Contracts for ordnance material were let by competitive bidding, and the estimates of the manufacturing arsenals were by no means invariably the lowest.

The manufacturing arsenals played a key role in the work of the Ordnance Department.[2] Not only did they perform important developmental functions in armaments such as were carried out in Europe by firms like Skoda, Krupp, and Vickers; they also provided the department with a yardstick for checking the reasonableness of the prices charged for ordnance material by private firms. If private bids for a contract seemed high, the department had in reserve the possibility of manufacturing the needed material itself. In addition, the arsenals performed a useful training function for officers of the department, and they served as valuable reserve productive capacity for unexpected requirements — a role they played after 1914, when orders from Britain and France began to swamp the facilities of private arms manufacturers. Besides all this, there was among the officers of the department a distinct feeling of pride in the arsenals, in the quality of work they turned out, and in the abilities of the skilled men who worked in them. The manufacturing arsenals were part of the valued tradition of the department.

When, therefore, Crozier appeared before committees of Congress to give testimony as to the conduct of his department and its need for funds, it was with more than routine earnestness that he argued for the maintenance and modernization of his manufacturing arsenals. His task was no easy one, despite the fact that the need for funds was, from the departmental point of view, urgent. None of the arsenals was up to date in its equipment or methods. In most of them the buildings and plant dated largely from Civil War times. Funds for capital equipment and rebuilding that a private concern would have secured from plowed-back profits or the capital market the arsenals had to obtain entirely from congressional appropriations. In years of peace Congress was not inclined to be generous. The problem that Crozier faced was not merely to secure funds for ordnance but funds that could be spent on the arsenals, rather than on purchases from private firms. And here prejudices had to be met. There was a general belief that the arsenals were less efficient than private firms — a belief based not on any informed comparisons of costs of production, for these were generally not to be had, but on an ill-defined impression that government manufacture must be inefficient. Congressmen did not *know* that the arsenals were mismanaged, but they believed that they probably were and that the burden of proof was on those who alleged they were not. Added to this was a certain bias in favor of giving the government's business to private firms whenever possible. Government arms contracts frequently involved large sums of money, and there was an understandable predilection on the part of Congressmen to channel such contracts into the hands of civilian manufacturers. Why should the taxpayer subsidize government competition with private industry? Apart from any political pressure the private iron and steel concerns might bring to bear, there was antipathy to the idea of the government as a manufacturer. If the arsenals could underbid private firms, it was unfair competition. If they could not, they were inefficient.

In attempting to defend the interests of his department against these prejudices, Crozier had two tasks: to obtain defensible estimates of the actual costs of manufacture of specific products at the arsenals, so that direct comparisons could be made with the contract prices of private manufacturers; and to take some positive action which would enable him to claim that the management of the arsenals was at least as efficient as that of private business. Crozier accepted the proposition that the arsenals *should* be as efficient as private firms, and he interpreted this to mean that their costs should be no higher than the contract prices paid to private firms. These two statements did not necessarily mean quite the same thing. The arsenals did a great deal of experimental work, they were responsible for the designing and testing of new weapons and equipment, and their production runs were typically much shorter than those of private arms manufacturers. If Crozier had wished to argue that slightly higher costs were justified at the government arsenals, he could have made out a good case. But he did not take this point of view.

Comparisons of efficiency between the arsenals and private manufacturers of armaments presented a number of problems.[3] Merely to compute costs of production for the specific products which the arsenals manufactured was no simple task, largely because of the primitive cost-accounting methods used. To compare these costs with the costs of private firms manufacturing different "mixes" of products and using different cost-accounting methods was almost impossible. Even when figures were available, it was a question debated frequently in congressional hearings whether or not any given set of cost figures provided by the arsenals really included all the items of cost which a private manufacturer had to bear. And even when this question was decided, it remained uncertain whether or not the arsenals had "unfair" advantages. Their executive personnel were relatively cheap. The commanding officer at Watertown, for example, cost the government less than $6500 a year, including his pay, living quarters, and all perquisites. According to Crozier, a man holding a post of comparable responsibility in private industry would have cost his employer at least twice as much.[4] Then again, the Ordnance Department, in figuring its costs, could calculate interest at 3 per cent, taking advantage of the government's better credit, whereas private manufacturers had to pay at least 5 to 6 per cent. And finally, government establishments incurred no costs for fire insurance, but only for actual fire losses; averaged out over all government operations, costs for actual fire losses, according to Crozier, amounted to only about one half the costs of fire insurance premiums for private firms.

To offset these advantages, the arsenals seemed to be under a handicap in regard to labor. The fact that the arsenals worked an eight-hour day, as against the nine- or ten-hour day of most private employers of the time, may or may not have resulted in lower productivity per man-year. As difficult to estimate were the effects of fifteen days' leave of absence with pay annually for each employee, seven national holidays annually with pay, and thirteen half holidays with pay annually during the months of July, August, and September. These generous vacations meant that the arsenals paid their employees for twenty-eight and a half days annually during which they did no work. This may not have been a net handicap in competition with private firms, however, because of the favorable effect on morale and because it enabled the arsenals to get the pick of the available skilled labor. Crozier personally was inclined to believe that the arsenals were in a position to skim the cream from the skilled labor reserves in their neighborhoods. Certainly they attracted men who wanted relative security of employment, average pay, and congenial working conditions. On the other hand, manufacturers in the vicinity of Rock Island Arsenal assured Crozier that they would never hire men who had been employed at the arsenal because it was "under the control of the unions, and the best men did not want to go there." [5]

A more important effect of the fringe benefits available to employees at the arsenals may have been to reduce labor turnover. Wage rates at the arsenals were set by civil service regulations at a level equal to the going rates for similar work in private firms in the vicinity of each arsenal, so that there can seldom have been any great financial inducement for a man employed at an arsenal to quit his job and seek work elsewhere, unless he was prepared to move out of the neighborhood entirely.[6] The better working conditions to be found at the arsenals furnished a positive inducement to stay. Among the skilled tradesmen, at least, turnover was very low. At Watertown Arsenal, for example, in May 1914 almost half the workers (49.77 per cent) had been employed continuously at that arsenal for four years or more. The average length of service for those who had been at the arsenal for at least a year was eight years.[7] If the floating population of unskilled laborers could be excluded from the calculation, turnover among the skilled employees would be low by any ordinary standards.

A definite saving in costs must have resulted; the expense involved in high turnover rates is well known. But the effect of good conditions on worker productivity is another matter. No doubt the employees at the arsenals were aware that their jobs were pleasanter than they would have been in private industry. Whether this led them to work harder or more efficiently is doubtful. A job at an arsenal was considerably more secure than a job in private industry. There was no danger of the firm going into bankruptcy. The number of jobs open varied with the orders on hand and with congressional appropriations, but the commanding officers were reluctant to lay off any of their skilled workers. A man could be fired for inefficiency or disobedience, but not just on the say-so of his foreman. Grievances could be and were taken to the officer in charge of the shop, from him to the commanding officer, and from him to the Chief of Ordnance —not to mention the ever-present possibility of an appeal to the local Congressman. General wage cuts were ruled out by civil service regulations, unless the level of wages in the vicinity of the arsenal also fell. But this security does not necessarily mean that productivity per man was higher than it would otherwise have been. Especially at the arsenals such as Watertown, where piece rates were not used, there was no incentive for the men to compete with each other in terms of output. The effect of good working conditions and job security was not to generate a steady upward pressure on production rates, but to encourage a work pace that could be maintained comfortably from day to day and which did not distress the slower, older, or less efficient workers.

Crozier's Annual Reports in the years between 1901 and 1915 contained repeated discussions of comparative costs. In general he was able to demonstrate, to his own satisfaction at least, that at the arsenals where repetitive work was done — where long runs of a standard product were feasible — costs of production were usually equal to the contract prices paid to private firms and occasionally lower. In 1911 small-arms ammunition was costing the government $26.95 per thousand rounds when manufactured at the Frankford Arsenal, but between $34 and $35 when bought on contract from private manufacturers. Cannon powder manufactured at Picatinny Arsenal cost 56 cents a pound in that year; when bought from private sources it cost 60 cents a pound.[8] But at the arsenals where the work was more varied and where long production runs were not possible, no such favorable comparisons could be made. This was particularly true of Watertown Arsenal. A single twelve-inch disappearing gun carriage, for example, cost the government $56,987.18 when manufactured at Watertown in 1906–1907; when purchased by contract in September 1907 the price was only $51,062.15.[9] In 1912 Crozier could assert before a congressional committee, "I have considered that the management of the arsenals has been good as compared with the management of the ordinary industrial establishments of the country."[10] But before the same committee he was compelled to admit, "For some time I have found that the cost at Watertown Arsenal compared less favorably with the cost of the same material procured from private manufacturers than at any other arsenal. It has made often the poorest showing." [11]

When Crozier undertook to compare costs at Watertown with prices paid to private firms, the firms he had to refer to were the Bethlehem Steel Company and the Midvale Steel Company, since these were the only concerns which manufactured the same types of gun carriage as did Watertown.[12] Both Bethlehem and Midvale could usually bid for contracts at prices lower than Watertown could match, even though their bids included a profit margin while Watertown's figure did not. Both, as Crozier was aware, had at one time made use of Frederick Taylor's services.

Precedents for the employment of civilian experts in the management of the arsenals, if Crozier sought them, were not hard to find. In 1815 Eli Whitney, pioneer of mass production, had been called upon to advise the Ordnance Department on the design and manufacture of a standard musket; and one of Whitney's mechanics, Roswell Lee, had been employed at Springfield Armory in the same year to introduce a system of managing the armory's two hundred and fifty employees — a system which, in words which Taylor or Barth might have used a hundred years later, could be counted upon "to provide stock, tools and materials for keeping them employed, to preserve order, subordination and regularity of exertion [and] to retain every branch of the business in a relative state of progression with the rest." [13]

Crozier had special reason to seek civilian aid, for his department was, in the years before 1909, very short of officers. The principal source of officer personnel for the Ordnance Department was normally the artillery regiments. From 1873 until 1901 the lowest rank in the Ordnance Department was first lieutenant, so that a second lieutenant of artillery who was accepted by the Ordnance Department received an automatic promotion — a lure which, in the slow years of peace, was not to be despised. Officers were selected by an examination which, though not required by law to be competitive, actually became so because the number of those who sought acceptance by the department was always greater than the number of vacancies. This examination, combined with the provision for automatic promotion upon entry, was, according to Crozier, successful in providing the Ordnance Department with a steady supply of high-quality officer personnel.[14]

After the conclusion of the Spanish-American War in 1898, severe criticism was directed against the staff departments of the Army on the grounds that they were too completely divorced from the line regiments and lacked knowledge of and sympathy with the requirements of the troops in the field. The cause of this was alleged to lie in the fact that there was no exchange of officers between staff departments and line regiments. Once an officer was admitted to, say, the Ordnance Department, he remained in that department for the rest of his Army career. To remedy this situation, an act was passed in 1901 which placed all staff departments (except the Judge Advocate General's, the Medical Department, and the Engineers) on a detail system; officers were detached from the line for four-year tours of duty in staff departments, with compulsory intervals of two years between successive details for all ranks below lieutenant colonel.

This new system may have remedied the alleged divorce between staff departments and line regiments, but it came near to wrecking the Ordnance Department's method of securing new officers. Officers detailed to staff departments now received no automatic increase in rank; consequently there was no incentive to seek service in a staff department except a strong liking for the work. Unfortunately, this incentive was not sufficient to secure officers for the Ordnance Department. The examination before entry was retained, but there were not enough applicants to fill the vacancies, and in some years there were no applicants at all.[15]

Crozier, upon his appointment as Chief of Ordnance in

1901, realized the damage which this system was certain to do to his department and immediately pressed for a revision. It took him five years to secure the changes he wanted. In 1906 an act was passed authorizing the transfer to the Ordnance Department (not to other staff departments) of officers from the same grade in other branches of the service *or* from the grade below, cutting the compulsory interval between successive details from two years to one, and lowering the grade at which the compulsory interval should cease from lieutenant colonel to major. This change placed the officers of the Ordnance Department on a competitive basis, both for entry and for subsequent service, with the strong incentive of a promotion upon either the first detail or some later one. Any officer who did not make good was not detailed to the Ordnance Department a second time. This system proved successful and was continued until the United States entered World War I.[16]

Thus, for five years after 1901, until the law was changed in 1906, the flow of new officers into the Ordnance Department was seriously reduced. Those who entered the department in this period did so only because they were strongly attracted by the type of work which the department carried out. There was no other incentive, and the entrance examination, the standards of which were maintained in spite of the inadequate number of applicants, presented a hurdle which only the determined would take the trouble to surmount. The total number of officers in the department during this period was always below the number (seventy-one) prescribed by the Act of 1901.

This acute shortage may have predisposed the Chief of Ordnance to favor a system of management which promised to dispense with the need for personal, piecemeal supervision of production in the arsenals. One attractive feature of the Taylor system was that it offered a means of extending the range and effectiveness of managerial control without a proportionate increase in the size of the managerial force; it was a means of economizing managerial manpower. This was an urgent matter for the Ordnance Department. Over the whole period from 1901 to 1913 the department was badly understaffed in its middle management echelons. As early as 1904 Crozier stated bluntly in his Annual Report that "the Department can not do its work with its present force of officers" and alleged that the quality and quantity of work were falling in consequence. He warned that, if measures were not taken to supplement the number of officers in the department, civilian engineers would have to be employed or the volume of work given out on contract to private manufacturers would have to be increased to create "a properly skilled corps of experts." And in 1905 he pointedly remarked that the strain of overwork was beginning to tell on the officers in the department, eight of whom out of a total of fifty-nine had been seriously ill during the year. In 1906 he began his negotiations with Taylor.

The difficulty of attracting officer personnel into the Ordnance Department also affected the decision to adopt the Taylor system in a different way. The timing of Crozier's efforts to get the Act of 1901 amended exercised a significant influence upon the timing of the introduction of Taylor methods at Watertown. The precise dates are important. Under the revised Act of 1906, a man who graduated from West Point spent his first two years as an officer with troops in the field. If he wished to enter the Ordnance Department, he was transferred at the earliest at the end of his second year after graduation. He then went for one year to Sandy Hook, the Ordnance Department proving ground. Having already received a grounding in mechanical engineering at West Point, he spent his year at Sandy Hook studying the design and operation of various types of ordnance material. At the end of this year, his third after graduation, he was available for transfer to one of the manufacturing arsenals.[17]

It is clear that the effects of the change which Crozier secured by the Act of 1906, which once again provided the Ordnance Department with a predictable supply of junior officers, would not be evident *at the arsenals* until 1909 at the earliest. Now, 1906 was the year when Crozier first approached Taylor, and 1909 was the year when Carl Barth began work at Watertown. This may be more than coincidence. In 1906, for the first time since his appointment as Chief of Ordnance, Crozier was assured of a supply of young officers. In 1909 these officers were ready to enter the manufacturing arsenals. Crozier had fought to get the Act of 1901 modified for good reasons; but having got his young officers, he had to train them. One reason for hiring Barth, which may explain the delay between Crozier's first approach to Taylor in 1906 and the final agreement in 1909, was that Crozier intended his first crop of officers admitted under the new regulations to receive on-the-job training under the most advanced system of management then available: the Taylor system.

The organization and functioning of the Ordnance Department during these years bore the clear imprint of Crozier's personality and policies. His role as Chief of Ordnance involved him in two principal sets of relations: with Congress and with his subordinates in the department. There were also his relations to his superiors in the chain of command, the Secretary of War and the President, but these individuals play no conspicuous part in our story. This contrasts sharply with the situation in the Navy Department, where it was primarily because of the opposition of the Secretary of the Navy that Taylor, despite continued efforts, failed to get his methods officially adopted in the Navy yards.[18] In the Ordnance Department Crozier seems to have been given a remarkably free hand by his superiors, at least as far as the manufacturing arsenals were concerned. The source from which he expected and experienced limitations on his freedom of action was not the executive branch of the government but the legislative, in particular the committees of the House and Senate which, in the course of approving the annual appropriations, examined in detail the functioning and policies of the ordnance and other military departments.

The general impression one gathers from reading Crozier's correspondence is that he was far from being a sheep among the wolves of Washington politics. Political sophistication was a vital requirement for anyone holding the job of Chief of Ordnance in this period, and Crozier's tenure of that office was long and distinguished. Insofar as the Taylor system is concerned, the political overtones of Crozier's policy are evident. The Taylor system was adopted by the Ordnance Department primarily as a result of congressional criticism of costs of production at the arsenals. It was finally thrown out of the department as a result of congressional action to outlaw time study, a direct consequence of pressure by organized labor on members of Congress. Crozier, as Chief of Ordnance, stood at the point where these political pressures centered.

Part of Crozier's job was to screen his subordinates from political interference. In particular, he stood between Congress and the commanding officers of the manufacturing arsenals. An excellent administrator himself, he was fortunate in having, to command these arsenals, a group of men who were thoroughly loyal to him and to the department, who carried out his orders promptly and efficiently, but who felt little constraint in criticizing his proposals before they became orders. This criticism, however, was always kept within the department; it never involved reference or appeal over Crozier's head. It was the kind of criticism, uninhibited in degree but delimited in range of participants, typical of a highly effective system of executive communication.

The commanding officers of the manufacturing arsenals during the period when the Taylor system was in process of adoption were Lieutenant Colonel Frank E. Hobbs at Rock Island (replaced by Lieutenant Colonel G. W. Burr in 1912), Colonel S. E. Blunt at Springfield (replaced by Lieutenant Colonel W. S. Peirce in 1913), Major George Montgomery at Frankford, Lieutenant Colonel W. W. Gibson at Watervliet, and Major C. B. Wheeler at Watertown. These men, with General Crozier and Major C. C. Williams, the officer who was detailed to work with Barth at Watertown, formed the group whose verdict on the practicability and utility of the Taylor system determined whether or not the department would attempt to retain it as its standard method of shop management. None of the group is on record as having opposed its adoption or as having been unconvinced of its utility after it was adopted. Williams, the officer most intimately involved in the installation of Taylor methods at Watertown, later rose to the top position in the department. After having been chief ordnance officer to the American Expeditionary Force in World War I, Williams succeeded Crozier as Chief of Ordnance in 1918, retaining that post until June 1930.

The officers of the Ordnance Department received no more formal training in management than did any American businessmen in this period — which is to say, none at all. The officers selected for managerial posts at the arsenals were not differentiated from other ordnance officers by special training or experience in management. Wheeler, for example, was probably chosen as commanding officer at Watertown because he was an expert on gun-carriage construction, gun carriages being Watertown's major product. Nor were managerial abilities, potential or actual, as distinct from "officer-like qualities," taken into consideration. The competitive examination taken by all entrants emphasized technical knowledge of artillery, mechanics, mathematics, and similar subjects. Crozier, describing what he meant by a good ordnance officer, stressed engineering and inventive ability. "The designing and constructing ordnance officer," he wrote,

. . . must be a mechanical engineer, since no characteristic of this mechanical age is more pronounced than the complete entry of its spirit into the production of implements and engines of war. The ordnance officer's knowledge of engineering subjects must not be merely that of the liberally educated man . . . but that of the expert with details at his finger ends.[19]

The role of commanding officer of a manufacturing arsenal was different from the role of private businessman, both in the functions performed and in the relationship of the individual to his job. Wheeler at Watertown had no need to concern himself with problems of profits, credit, pricing, advertising, or sales. Other functions were so transformed as to be noncomparable. Capital supply, for instance, was a matter of congressional appropriations, not of offering shares or bonds on the money market. On the other hand, the commanding officer of a manufacturing arsenal was confronted with precisely the same problems of product design, productivity, costs, and shop management as was the production manager of any other large metalworking establishment. To make a distinction which is to some extent artificial, the financial functions of the private business executive were irrelevant to the job of the commanding officer of an arsenal, while the production functions were identical. Because his attention was not distracted by problems of finance and marketing, the ordnance officer in charge of an arsenal was likely to attach greater importance to shop management and productive efficiency.

The commanding officer of a manufacturing arsenal must be thought of as a salaried manager. He held his position not by ownership rights but by right of appointment and rank in the departmental hierarchy. This implied that his interest in doing his job well stemmed not from any expectation of direct pecuniary gain but from the hope of advancement in his career, from loyalty to the organization, and from the expectation of winning recognition from his peers and superiors.

For men like Wheeler and Williams these motives were apparently more than adequate; in the period before World War I, any competent ordnance officer — and the department was not slow to rid itself of the incompetent — could take his pick among offers of employment at much higher salary from private business firms, particularly those engaged in armaments manufacture and those concerned with government contracts. Resignations from the department to accept such offers explain the serious drop in officer personnel in 1915, when private armaments manufacturing firms began expanding to handle orders from Britain and France.[20] Officers who remained in the department in these years can have been induced to do so only by a strong sense of identification with the organization or by the high probability of rapid promotion.

All commanding officers of manufacturing arsenals had certain characteristics in common. They were trained mechanical engineers and were selected primarily on the basis of technical knowledge and competence. Their jobs involved them principally in production management functions. Perhaps partly because of their technical training, partly because during their formative years they had been thoroughly indoctrinated with military discipline, they tended to show a certain insensitivity to human relationships, at least in dealing with their employees, and to interpret work situations in authoritarian terms. One incident not directly connected with the Taylor system may illustrate this insensitiveness. In the years before 1908, Watertown Arsenal was acquiring a reputation in the Ordnance Department as the one arsenal at which costs of production were inexcusably out of line with the costs of private arms manufacturers. This was the reason why Wheeler — the new broom — was appointed to Watertown in 1908 and why Watertown was the first establishment selected for the installation of the Taylor system. In June 1907, Lieutenant Colonel Hobbs, Wheeler's predecessor, reported to Crozier on the morale and work habits of the Watertown employees. He found little to criticize. "Generally speaking," he wrote,

. . . the employees are an efficient and trustworthy body of men. In visiting numerous private establishments in this vicinity, and then *in making occasional inspections of the Shops in civilian dress,* I have been struck by the fact that there seems to be much less idleness or talking or visiting in the Shops here than in the private establishments.[21]

Hobbs was one of the most hard-working and conscientious officers in the department, a man whose initiative is demonstrated by the fact that he introduced time study at Rock Island Arsenal, without orders from Crozier, before it was tried at Watertown. Yet in this report he takes it for granted that by changing out of uniform into civilian clothes he can walk through the workshops at Watertown without being recognized as the commanding officer.

Nevertheless, ingenuous as they may have been in interpreting their relations with their subordinates, the officers of the Ordnance Department showed a degree of sophistication in other respects. They were immeasurably more sensitive than any of the Taylor group to the possible political repercussions of an attempt to install Taylor methods in the manufacturing arsenals. Crozier, Wheeler, and Hobbs were far more concerned about the possibility of exciting the hostility of their employees and precipitating congressional action than were Taylor and Barth. Time and again in his early letters to Taylor, Crozier expressed anxiety lest the introduction of Taylor methods might cause resentment among the workers and lead to protests to local Congressmen, only to have his fears ridiculed by Taylor as unworthy of serious consideration. When Crozier first explicitly laid out his problem for Taylor's advice, he phrased it in political terms: "My principal difficulty now is in arranging such methods of payment of workmen as will not result in pressure by the Unions upon

Members of Congress for change, which the Members will find it difficult to resist." [22] And again: "I am hoping for some suggestions which may make it possible for us to establish proper rates at the Arsenal . . . without stirring up apprehension which can easily become bad feeling, which in a government establishment always endeavors to find its outlet through the local Member of Congress, who is thus likely to be placed in a difficult situation." [23] When Dwight Merrick, Carl Barth's assistant, went to Watertown to begin time studies, Crozier seized the opportunity to stress once again the danger of political action: "You know that I am relying upon Mr. Merrick's experience with you [Taylor] to insure his proceeding with tact, and the continuation of the good relations existing at Watertown Arsenal and the consequent absence of appeals to local Congressmen which they are not in a position to neglect." [24] When trouble broke out at Rock Island Arsenal over Lieutenant Colonel Hobbs's ill-considered attempts to start time study, the warning was underlined: "The workmen . . . took the method usual in government establishments of appealing to Congressmen. As the workmen at the Rock Island Arsenal represent a good many voters they had little difficulty in enlisting the efforts of the Congressmen, and this condition must always be looked for. It is impossible for the Congressman in a district in which there are a good many government employees to ignore the sentiments of such voters; and the sentiment is apt to be most prominently represented by the unions." [25]

What Crozier was trying to convey to Taylor was that the workers at the government arsenals had an effective and well-institutionalized means of defending themselves against any changes in the conditions of work which they regarded unfavorably. But Taylor refused to take the warning seriously. "I do not think," he wrote to Barth in April 1910, "that there is any occasion to bother much about what the American Federation of Labor write concerning our system. I do not believe that they can seriously interfere with it. The important matter is to keep the people straight at Washington." [26] And in March 1911, just before the strike at Watertown, he wrote confidently to Crozier, "I have no idea you will entirely escape correspondence from members of Labor Unions, but do not believe that it will amount to much." [27]

This cavalier attitude toward the possibility of labor trouble can in part be explained in terms of Taylor's previous experience in introducing his methods into private manufacturing plants. Until 1912 they had never been introduced into any concern where the labor force was strongly unionized or where a union was recognized as a bargaining agent. There had been a strike at Sayles' Bleacheries while H. L. Gantt was working there, but it was not a strike against the Taylor system. In February 1909 there was a strike at Bethlehem, where Taylor had installed some of his methods between 1898 and 1901, but here again — as Taylor explained it — it was not a strike against the Taylor system but rather against Schwab's attempt to get more work out of the men without giving them more pay.[28] In other plants where Taylor methods had been adopted, individual workers who belonged to unions had quit either their union or their jobs. If they quit their union, that was a victory for the Taylor system. If they quit their jobs, their places could easily be filled either by hiring new men or — preferably — by training unskilled and therefore cheaper labor to perform parts of the jobs which the skilled workers had been doing. The hostile attitude of the craft unions, up to the time of the molders' strike at Watertown, was not something about which Taylor had been forced to worry very much.

The essential difference between the situation at Watertown Arsenal and the situation at the private firms where Taylor methods had previously been installed lay not in the number of employees who were union men but in the ability of the employees, and of the craft unions in the name of the employees, to take political action for redress of their grievances. Neither Crozier nor Taylor seems to have considered the possibility that there might actually be serious labor trouble at Watertown; the reluctance of organized labor to strike against the government was well known. But Crozier realized from the first, as Taylor did not, that the workmen at the arsenals, without relying on the strike weapon, could make their influence felt effectively through political channels and were equipped to resist any changes in their working conditions which they regarded as not in their interests.

The man who showed the most acute appreciation of labor's probable reaction to attempts to install Taylor methods in the arsenals was, interestingly enough, Lieutenant Colonel Hobbs — he who in 1907 had evinced such faith in the ability of civilian dress to disguise his identity. Hobbs was an old friend of Taylor's and had followed the development of the Taylor system since its beginnings at Midvale. Acting on his own initiative, he had attempted in 1908 to set piece rates at Rock Island Arsenal by means of stop-watch time study. The attempt had failed because of the intense opposition it aroused among the workmen, who had appealed immediately to their local Congressmen.[29] Hobbs seems to have learned a great deal from this experience. Writing to Crozier in February 1909, he warned him not to be deceived by Taylor's easy optimism:

I am sure there will be some opposition to the introduction of the system anywhere. . . . It would be interesting to know just what the trade organization conditions are in and near the places where Taylor has operated. I feel pretty sure that in some there has been no such thing in existence and in others there has been a very large labor market close at hand on which to draw.[30]

And, undeterred by his earlier failure, he proceeded to argue that Rock Island Arsenal should be the first establishment to be reorganized under the Taylor system, not because labor conditions there were easier than elsewhere but because they were more difficult!

You speak of possibly applying Taylor's system either at Watertown or Frankford. Watertown is comparatively small and Frankford employs a great many girls, both are near very large and practically inexhaustible labor markets. If you want to know how the system is really going to work out in a government establishment apply it here where the conditions, generally speaking, are the worst and hardest, where there are simple operations to learn on and complicated operations to work up to later. If this can be done here there will be no question about the other places.[31]

The argument did not appeal to Crozier. Conscious that he was innovating, Crozier decided to conduct his experiment in the plant where the chances of success, regarding labor conditions, seemed greatest, not least. If the Taylor system worked at Watertown, it could be extended to the other arsenals later.

Crozier's decision to select Watertown Arsenal for his initial attack on the problem of high production costs was not taken solely because of its proximity to a large labor market. Of all the arsenals, Watertown was most conspicuous for its apparent inability to match the costs of private manufacturers. To some extent this was due to the wide variety of products that Watertown turned out. Its principal product was seacoast gun carriages — large and complicated mechanisms, involving the manufacture and fitting of some 4500 different components. Orders for these gun carriages were usually for small numbers, not more than four or five at a time, and each carriage took several months to build. A single order involving the expenditure of some $100,000 would necessitate, therefore, the manufacture of no more than half a dozen each of several thousand different components. There was, in consequence, practically no opportunity for continuous runs on a standard product. The arsenal also manufactured a wide variety of smaller articles, such as gun forgings for mobile artillery, frames for packsaddles, experimental models of a new type of aluminum water bottle, artillery targets, and armor-piercing projectiles for seacoast guns. These were usually manufactured in much larger quantities than the gun carriages, but once an order was filled there was no assurance that it would be repeated.

Watertown Arsenal, then, was an establishment where large-scale engineering work and small-scale jobbing work went on side by side. The normal situation was for the arsenal to be working on a great variety of products, manufacturing small batches of the larger and more complicated items and limited quantities of the smaller items on special order. This made it difficult to set up regular production sequences; it also made straight piecework impossible, because jobs had to be changed so frequently.

For several years prior to 1909 the Annual Reports of the Chief of Ordnance carried complaints of inadequate and obsolescent productive equipment at Watertown. In 1907 and again in 1908 Crozier seized the opportunity to point out that the machine shop was handicapped by a large number of old machines not adapted for the use of the new high-speed steel, by inadequate electric generating capacity, and by a lack of facilities for handling heavy work. Conditions in the smith shop and foundry of this arsenal, he asserted, were even more unsatisfactory, the equipment in these shops being largely obsolete and entirely inadequate for the work required of it.[32]

Early in February 1908, Major (soon to be Colonel) Wheeler was appointed commanding officer of the arsenal and immediately entered upon an energetic campaign to lower costs and increase productivity.[33] While Crozier, with his eye on congressional appropriations, had emphasized obsolete and inadequate capital equipment as the cause of

Watertown's poor record, Wheeler was inclined to lay greater stress on poor shop management. In his correspondence with Crozier he was emphatic on the subject, in a way which would hardly have been possible if he had been longer in command of the arsenal. "There is at present practically an entire absence of shop methods here," he wrote in November 1908, "and it is only necessary to make a visit to and inspection of the shops to realize the importance of and necessity for a change." [34] He proceeded to recommend certain indispensable reforms: a toolroom with the necessary facilities for forging high-speed steel, the installation of a belt department, and the alteration of certain machines to give them sufficient power to enable high-speed steel to be used efficiently. In January 1909, he wrote, "I hesitate to describe fully the deficiencies of this Arsenal, because they are difficult to understand by those not in close contact with it. There is, however, an immediate need of a system that will enable us to turn out work efficiently and at moderate cost. This is not now being accomplished, notwithstanding all the urging which has been given." [35]

But Wheeler was not content merely to complain to his superiors and then wash his hands of the matter. In the twelve months which followed his appointment to the arsenal he instituted a series of important reforms. He reorganized the costing system by establishing separate shop expense ratios for each department, instead of the single over-all shop expense ratio which formerly had been applied throughout the arsenal. He installed a system of electric time clocks and job cards, by which the workmen stamped on their cards the times of beginning and ending each job. He began to keep the costs of individual jobs. [36] He ran experiments with high-speed tools and disciplined two machinists who refused to operate their machines at the speed he prescribed. And he set up an engineering division, whose function it became to estimate material requirements and costs, to carry out detailed analyses of orders, to issue instructions to the shops regarding the work to be done, and in short to consolidate in one department many of the functions which had previously been performed by the foremen.

But this attack on the problem left Wheeler dissatisfied. Anxious as he was to produce quick results, he became convinced that no substantial reductions in costs could be achieved without instituting some form of premium or incentive wage system. But this was easier said than done; piece rates of the ordinary type were not feasible because of the absence of repetitive work. In November 1908, therefore, he suggested to his superiors that the Taylor system, "with certain modifications which would be necessary in a government shop," was the one most worthy of adoption.[37] And in January 1909, in a long letter to General Crozier, he argued that "The adoption of his [Taylor's] system by the Dept. and its introduction at this arsenal where the necessity for it is most urgent, would mark one of the greatest strides in advance in recent years." [38]

To support his argument, Wheeler described in detail a recent job in the Watertown foundry which, he said, illustrated the inefficiency of all the shops. Certain employees in the foundry had been engaged in molding side frames for twelve-inch mortar carriages; their rate of production was one every five days. After several side frames had been produced, Wheeler became convinced that this rate was ridiculously low. He therefore canceled the work order and gave the remainder of it out on contract to private manufacturers. At the same time he asked Colonel Peirce, an ordnance inspector, to report on the rate of production in private foundries. The comparison was startling. At the Federal Steel Foundry Company Peirce reported that two molders and a helper produced a pair of these same frames in one day; at the American Steel Foundry Company two molders, one helper, and an apprentice, "if they are hustled," could turn out a pair and a half in a day. Wheeler commented:

At the Federal Steel Foundry the cost to mold one side frame at our rates of pay is only $3.92. Compare this with our cost, $24.20!! Yet the molders here *complain* "that they are working at a piece work pace for a day's wage" and some want their pay increased.[39]

The changes he was making would in time do something to remedy these anomalies, but it would be a slow process. Taylor's system, on the other hand, "has already been developed and we need it, and it will have hearty support. . . . I think, therefore, we ought to plunge right into his system and introduce it as quickly as possible." He concluded by recommending that Crozier should communicate with Taylor at once.[40]

Wheeler could hardly have been as explicit as he was in recommending the adoption of the Taylor system of management if he had not been aware that some such step was already under consideration by his superiors. The possibility of adopting Taylor's methods, and of hiring either Taylor or one of his associates as a consultant, had been in Crozier's mind for at least three years. Nor was the proposal as radical as it seemed. To some of the officers of the department Taylor was no stranger and his methods far from unknown. Three of them, Colonels Hobbs and McNutt and Captain Ruggles, had known him at Midvale and at Bethlehem, where they had been stationed as ordnance inspectors. Through Ruggles, Taylor, while still at Bethlehem, had been introduced to Wheeler.[41]

Negotiations between General Crozier and Taylor began in December 1906. Early in that year Taylor had been elected president of the American Society of Mechanical Engineers. In that capacity he was invited to attend a series of tests held at Sandy Hook, the Ordnance Department proving ground, and there he renewed an acquaintanceship with Crozier. The opportunity to advance the cause of the Taylor movement was not missed. On his return to Philadelphia Taylor sent Crozier some literature on the Taylor system, probably the papers he had earlier presented before the A.S.M.E. He added:

I have no doubt you will say when you receive the several pamphlets on Shop Management that I am sending you, "There goes Taylor riding his old hobby." However, I feel sure that in our type of management we are on the right track, and it would give me the greatest pleasure if you personally could spare half a day to come to Philadelphia to actually see what we are doing.[42]

The invitation was extended to Crozier's subordinates also.

The tone of Taylor's letter was not that of a man introducing a strange subject for the first time. Clearly he assumed that Crozier already knew what his "old hobby" was. Nor was Taylor's approach that of a salesman; the president of the A.S.M.E. had no need to play the huckster. If anything, he underplayed his hand, relied upon the prestige of the society to gain attention, and invited the pragmatic test of "come and see."

Crozier replied within four days. His letter makes clear that, like others who knew in a general way what Taylor was advocating, he thought of the Taylor system of management as centering around a system of incentive wages. He stated his belief that the officers of the Ordnance Department should become familiar with Taylor's ideas. Of the five manufacturing arsenals, he informed Taylor, piecework was used at Springfield and Frankford and a combination of piecework and a day wage system at Rock Island. At Watervliet and Watertown no piece rates were used.

As the work at Watertown and Watervliet is to a very little extent of the duplicating character we can not be said to have at either of those places anything better than the old time method of daily wages, with prospect of promotion by selection, and the occasional discharge of the poorest, to stimulate workmen to do at least a fair proportion of what they should be able to.[43]

In conclusion, he promised to come to Philadelphia and also to send Major Hobbs, then commanding officer at Watertown, Colonel McNutt, commanding officer at Watervliet, and Colonel Heath of Frankford Arsenal to visit Taylor.

Crozier did not keep this engagement, but the reason is unknown. For the next two years there seem to have been no written communications between him and Taylor on the subject of introducing Taylor methods into the department. In March 1907, however, Crozier sought Taylor's counsel on the best way to spend a slightly increased appropriation for a testing laboratory at Watertown and was advised to run a series of experiments on high-speed steel.[44] Clearly, therefore, the two men remained on terms of mutual respect.

Meanwhile the proposal to reform the management of the arsenals along Taylorist lines was carried forward by Crozier's subordinates. Hobbs wrote enthusiastically to Taylor as soon as he heard that Crozier had expressed interest, saying:

I am glad to know that the supreme authorities are possibly coming to your system of management. I have been preaching and proposing it in season and out of season, but had not had the opportunity to speak of it for nearly a year now, so I presume some other force must be responsible for the matter coming up at this time.[45]

And when Hobbs left Watertown to take over the command of Rock Island Arsenal, his successor, Major Wheeler, lost no time in getting in touch with Taylor and asking specific advice. His second-in-command, Major Ruggles, wrote to Taylor in February 1908 to explain that Wheeler was anxious to introduce a system of management similar to Taylor's at the arsenal but was concerned over the type of incentive wage plan to be used. What he wanted was a premium system, rather than a differential piece rate of the type Taylor usually advocated, because he wished to insure that each employee would never get less than his full daily wage. But above all it should not be a complicated system. "What we want is something as simple as possible, and that can be quickly introduced, as it is not desired to put in too elaborate a system at the start." [46]

More than a nodding acquaintance with Taylor's ideas was indicated by Ruggles' unquestioning assumption that even a simple system could not be introduced without time study. The letter asked whether Taylor could put him in touch with a man who had the proper instruments and data for setting task times. Taylor recommended that Ruggles should get in touch with H. L. Gantt or Carl Barth, but he also emphasized strongly that it would be useless merely to install an incentive wage plan. The Taylor system had to be introduced complete if the desired results were to be achieved. "Anything short of this leaves such a large part of the game in the hands of the workmen that it becomes largely a matter of whim or caprice on their part as to whether they will allow you to have any real results or not." [47] The goal was not simply the provision of incentives to which the workmen could respond or not as they chose; it was, ideally, control of the entire job situation, exercised in such a way that a predictable increase in output would certainly result.

By early in 1908, then, the chain of events which was to take Carl Barth and his assistant, Dwight Merrick, to Watertown had reached the stage where definite advice was being asked and definite recommendations made. Crozier knew that he was buying no panacea; nevertheless, as a calculated risk, the experiment seemed worth while. At the least it held out the prospect of reducing costs of production at the arsenals without the heavy expense that rebuilding and the purchase of new capital equipment would have involved. With apparently incontrovertible evidence of the superior efficiency of Bethlehem and Midvale before his eyes, with

Hobbs preaching the virtues of the Taylor system "in season and out of season," and with Wheeler, as soon as he arrived at Watertown, castigating the "entire absence of shop methods" which existed there, Crozier, in the deliberate and careful way which was characteristic of him, concluded that the interests of his department demanded the installation of Taylor methods. "It may require a little time," he wrote to Taylor in 1909, "for me to make a thorough examination . . . and to give effect to my intention to go into the subject; but I intend to keep at it." [48]

If there were good reasons why the Ordnance Department should be interested in the Taylor system, there were also good reasons why Taylor and his disciples should be anxious to have their methods adopted by the department. These went beyond the usual reasons for welcoming a new and important client. The government had a reputation as a good employer; it set the standards in industrial welfare and working conditions which private employers followed. The adoption of the Taylor system by the Ordnance Department would set an official seal of approval upon Taylor's aims and methods. Public knowledge that the department had examined his system of management and found it good would confer tremendous prestige upon Taylor and his movement.

The circumstances must have seemed unusually favorable. In certain matters, such as material procurement and wages, government regulations threatened to be a trifle confining. But relations with the permanent managerial force promised to be harmonious. Crozier was keenly interested. Wheeler, newly appointed to Watertown, could afford to be critical of the state of affairs which he found on his arrival and was not likely to resent the advice of persons whose employment he had himself recommended. The officers, secure in their status and military rank, were unlikely to regard the temporary employment of a civilian expert as a threat to their positions. Then again, Watertown was a metalworking establishment — the kind of plant in and for which Taylor methods had originally been devised. Problems of adapting the practices and standards already worked out were likely to be minimal. Full advantage could be taken of the cost-reducing potentialities of high-speed steel. And lastly, the job which had to be done was not too precisely specified. It was not a question of a man's being hired to install merely a new costing system or a new set of stores procedures; he was hired to reorganize the whole establishment insofar as government regulations permitted. This was the kind of opportunity which came seldom. At only two other firms — the Tabor and Link-Belt companies — had Taylor and his associates been given the same degree of freedom.

An awareness of these advantages — and perhaps of the splintering tendencies already evident within the ranks of his associates — was reflected in the care which Taylor took to see that a man whom he could trust and control was chosen for the work at Watertown. When Wheeler first took up with the Chief of Ordnance the question of introducing the Taylor system, the man he wanted for the job was Henry L. Gantt, who lived in Providence and who had already visited the arsenal more than once. In January 1909 Wheeler suggested that Crozier should get in touch with both Taylor and Gantt, saying that he was "under the impression" that they were still associated but was not certain.[49] In February he wrote, more positively, "In case we can enter into an agreement with Mr. Taylor, I think it would be greatly to our advantage if Mr. Gantt . . . could have direct supervision of the introduction of the system. He appears to me by far the most suitable man I know of for the purpose, and has on every occasion impressed me most favorably."[50] Crozier at first hoped that Taylor would undertake to supervise the work in person; when it was made clear to him that Taylor no longer accepted such assignments, he followed Wheeler's suggestion and asked for Gantt.

By this time, however, Taylor was determined that, whoever went to Watertown, it should not be Gantt. Gantt had been unwise enough to tell Taylor, early in 1908, what in his opinion could and could not be done at Watertown, and his prognosis was by no means to Taylor's taste. Gantt was convinced that it would be highly inadvisable to attempt to install the entire Taylor system at the arsenal: the great variety of jobbing work done there and the constant changes in product made thorough standardization impossible. The most that should be done, in his opinion, was to try and get a little more production out of the men by installing an incentive-payments scheme of the "task and bonus" type.[51] To Taylor this was heresy and doubly objectionable: first, because it implied that there were certain types of manufacturing to which the Taylor system was not applicable, and secondly, because it was a cardinal tenet of Taylor doctrine that the installation of an incentive-payments scheme without the rest of the Taylor methods was not merely ineffective but positively harmful. If Gantt went to Watertown, Taylor would be unable to share the credit without giving implied approval to Gantt's methods. Consequently, despite the fact that he had earlier recommended either Gantt or Barth as equally competent, Taylor now proceeded to cast doubt upon Gantt's qualifications and to propose in his place individuals who were free from heresy.

Naturally, Taylor was reluctant to alienate Gantt permanently. Writing to Crozier in January 1909, he advised him that it would be a mistake to employ Gantt, as he was "not a machine shop man," but at the same time asked that his advice be kept confidential:

It places me in a way in a very embarrassing position not to recommend Gantt, after he has already put in a certain amount of time and trouble on this same job, but I feel forced to do so. Please, therefore . . . do not refer in any way to anyone about my personal recommendation in this matter. You can readily see that Gantt might feel seriously offended with me under the circumstances.[52]

At the same time he wrote to Barth, enclosing a copy of this letter to Crozier:

I do not want Gantt to get mixed up with this, even although he has put in a considerable amount of his personal time in working up the business there. Therefore please do not mention to him at any time this letter. . . . I have written the Chief of Ordnance that you and Hathaway were the only two who would be competent to make a report at Watertown, and of the two that you had more experience than Hathaway.[53]

That Crozier should select Barth was inevitable. But it is intriguing to speculate what difference it might have made in the final outcome if Gantt and not Barth had been chosen. The two men were different, both in their personalities and in their understanding of what it meant to be a management expert. Gantt, in his relations with the permanent executives of the firms where he worked, was invariably diplomatic and prepared to compromise. He was willing to accept their definition of what needed to be done and of what he could and could not change. His attitude was that of the teacher, not the autocrat. He regarded his function as that of giving advice and instruction, not taking executive responsibility and along with it executive authority.[54] His statement that all that should be done at Watertown was the installation of an incentive-payments system almost certainly reflected not his conviction that this was all that was desirable but his belief that this was all that the arsenal officers really wanted. Gantt was prepared to depart from the rules of behavior which Taylor laid down and to follow his own conception of what his job required. Barth, on the other hand, accepted without question Taylor's intellectual leadership. His principal fault, on which he prided himself immoderately, was lack of tact. A first-generation immigrant, he affected a fine contempt for American technical education and machine-shop practice, permitting few opportunities for unfavorable comparisons with his native Sweden to pass unobserved. Frequently such comparisons took the form of immodest assertions of his own competence and ill-concealed contempt for the knowledge and abilities of his employers. Dogmatic, self-assertive, and obstinate, he was, nevertheless, a faithful disciple of Taylor in a sense in which Gantt, born on a farm in southern Maryland and a graduate of Johns Hopkins, never was.[55]

It is conceivable that there were situations in which Barth's blunt assertiveness was more effective than Gantt's diplomacy in getting done what needed to be done. But in analyzing what happened at Watertown, one must realize the kind of hostility and resentment that a man like Barth was capable, unintentionally, of arousing. The impression he made on some of the ordnance officers is illustrated by their reactions to his later work for the department, when the first luster of his prestige had dimmed. Writing to Wheeler in January 1918, Lieutenant Colonel George Montgomery stated that Barth, in spite of having received "every possible amount of co-operation" while introducing Taylor methods at Frankford Arsenal in 1910–1911, had "incurred the hostility of the men in the shops" and had so alienated the officers in charge that he "became quite impossible." Later, according to Montgomery, when Barth went to Rock Island Arsenal, his conduct became so intolerable that the commanding officer of the arsenal stated in confidence that "either he or Barth would have to go." Gantt, who worked at Frankford after Barth left, was warmly praised as "big brained and big hearted." [56]

Both Barth and Gantt were trained engineers and had originally become associated with Taylor in connection with his metal-cutting experiments. Gantt had helped Taylor at Midvale in 1887, while Barth had joined him at Bethlehem in 1899. Both men were thus fully conversant with machine-shop practice and with the new technology of high-speed steel. But Gantt had additional experience, which Barth lacked, with Taylor methods in foundries, for after leaving Bethlehem he had worked with the American Steel Casting Company and in the foundry of the American Locomotive Company at Schenectady.[57] This experience distinguished Gantt from all other Taylor consultants; he was the only one who knew anything about foundry practice and the problems of adapting the Taylor system to foundry work. Barth, when he went to work at Watertown, ignored the foundry entirely, despite the fact that it was the department which had been regarded by the arsenal officers as most in need of reform. The Taylor men at Watertown had no contact with the foundry until Merrick began to take time studies there, without any preparatory work, in the summer of 1911. It is hard to believe that this same ill-advised order of procedure would have been followed if Gantt had been in charge.

Perhaps, however, it would not have made any significant difference if Gantt had been in charge. One can argue that although they were poles apart in personality and temperament, Barth and Gantt would have interpreted the requirements of their job in much the same way. Both were engineers, both were in the consulting business for money, and both had experienced prolonged exposure to the ideas of Frederick Taylor. They necessarily shared many of the same prejudices and preconceptions. If Gantt would have acted at Watertown differently from the way Barth acted, would not the difference have been more a matter of pace, of sequence, of the personal phrasing of his role, than a substantial difference in what was attempted? The assumption on which this argument rests is that the conflict which later developed at Watertown had little to do with whether or not Barth was easy to get along with, but was a result of the painful disruption of work habits and shop organization which the installation of the Taylor system necessarily involved, no matter who was supervising the work.

Gantt had made several visits to Watertown in the years before 1909 and was on friendly terms with Wheeler and Williams. It appears probable that Wheeler's energetic attempts to convince Crozier of the necessity for installing the Taylor system reflected Gantt's influence, and the managerial reforms that Wheeler instituted at the arsenal during his first year as commanding officer may have stemmed from Gantt's suggestions. But this careful preparing of the ground was of no avail. Taylor was not willing to recommend Gantt, and Crozier would accept only a man who carried Taylor's seal of approval. The choice, therefore, narrowed down to Carl Barth or Horace Hathaway. Barth had the greater experience of the two and the greater reputation as a machine-shop expert. And so it was finally agreed. Taylor and Barth would go to Watertown, look the plant over, and report to Crozier what in their opinion could and should be done.

Chapter 3

THE ARSENAL

Watertown Arsenal is situated about eight miles west of downtown Boston, from which it can be reached without difficulty by streetcar or automobile. It presents today a somewhat impersonal face to the world, and the casual passer-by, peering through steel fences or staring at tall brick walls, sees few indications of the work that is going on within. Only occasionally do the arsenal and its work appear in the press. Once in recent years careful readers of the Boston papers may have learned that the gun carriages for the Army's much-publicized atomic cannon were made at Watertown, and there has from time to time been news of pioneering work on titanium alloy steels. But the arsenal seems today to function in a private world, meeting its own emergencies and celebrating its own achievements largely apart from the hurly-burly of business and politics. So it has been throughout most of the arsenal's long history.

But there have been periods when the arsenal was very much part of the American social drama, and its work intimately involved in politics and business. The years from 1909 to 1915 were such a period. At that time the arsenal covered an area of almost eighty-seven and a half acres, lying between Arsenal Street on the north and a leisurely curve of the Charles River on the south. It was made up of five main buildings — the smith shop, the machine shop, the erecting shop, the foundry, and the administration building — and employed, on the average, about five hundred civilians in addition to its military personnel.[1] By far the largest department, in terms of numbers employed, was the machine shop, with more than 50 per cent of the civilian work force. Next largest was the foundry, with just under 10 per cent. Funds expended at the arsenal for all purposes (maintenance and re-equipment as well as current orders) averaged, in the years between 1908 and 1913, around $800,000 annually.[2]

Technically, Watertown Arsenal in 1909 was far from an up-to-date manufacturing establishment. Much of the equipment was obsolete, by the standards of private business, and had suffered from unsystematic maintenance. In the machine shop about 40 per cent of the machine tools had been in service for fifteen years or more, and many of them for over twenty years.[3] All had been designed when carbon tool steel was the only type of tool steel available, and the system of belts and pulleys had been set up on that basis. Throughout the plant there was a serious shortage of small parts, such as bolts, straps, and clamps.[4] Handling facilities, such as cranes and trolleys, were inadequate.[5] The buildings were ill-lighted and cramped; the newest was fifty-three years old.[6] Overcrowding was a problem in every shop, but particularly in the foundry. In this department the only mechanical assistance available to the molders were two cranes and a few pneumatic rammers. For the rest, the work was entirely by hand.[7] Conditions in this shop were especially difficult because three different metals had to be melted, poured, and cast: bronze, cast iron, and steel. Bronze was poured practically every day, steel once or twice a week, and cast iron three or four times a month. On pouring days the rest of the work in the foundry was brought to a standstill, as experienced helpers were taken off their regular jobs.[8] The large variety of work done in the foundry also presented difficulties, but the inadequate handling and moving equipment and the interruption of regular work on pouring days seem to have been the most serious drawbacks.

Organizationally, the chain of authority at the arsenal ran from the commanding officer to his immediate subordinate, the officer in charge of the shops, from him to the master mechanic, and from him to the foremen of the individual departments.[9] There was a headquarters office staffed by clerks, whose principal function was correspondence, and an engineering division, an innovation introduced by Wheeler in 1908, which was responsible for the making of cost estimates, the preparing of blueprints, and similar tasks.[10] Managerial control was exercised chiefly by written instructions from the commanding officer and his aide, supplemented by verbal consultations with the master mechanic and the foremen. In the day-to-day operations of the arsenal the master mechanic and foremen were largely left to their own devices to allocate jobs and get out the work.

The system of cost accounting, particularly the method of allocating overheads, will be discussed in detail later, in connection with the reforms instituted after 1909. Up to 1906 the arsenal had allocated all costs directly to orders, with no attempt made to ensure that each product bore its appropriate share of the overhead; from 1906 to 1909 the standard "percentage on wages" method was used, each product bearing a share of the overheads in proportion to its direct labor costs. Stores procedures are best discussed in connection with the new methods introduced as part of the Taylor system. Purchasing methods were prescribed by statute and therefore were not subject to modification.

In April 1909, at the request of General Crozier and partly in response to Colonel Wheeler's urgings, Frederick W. Taylor and his associate, Carl G. Barth, visited Watertown Arsenal to investigate the possibility of installing the Taylor system of management.[11] Each submitted a report, Barth's being the more detailed since it was anticipated that he would have responsibility for the installation.[12]

Barth began his report by criticizing the way in which the arsenal's clerical work was handled. The number of clerks in the purchasing and property divisions was unreasonably high; much of the work was done in a roundabout and laborious manner; the bookkeeping methods were unnecessarily elaborate and cumbersome. He proposed, therefore, to reorganize the storeroom, the stores accounts, and the cost-keeping methods to secure a perpetual inventory of stock purchased and material on hand. Pointing out that until a short time previously the methods of shop management used at the arsenal had been of a primitive type, with most of the real tasks of management loaded on to a few foremen, he congratulated Wheeler on the reforms which he had recently instituted but argued that they could not be of any lasting or widespread benefit. This could be achieved, he asserted, only by a "complete reorganization of the whole shop management along the lines of the Taylor System."

At this point in his report Barth digressed to observe that Wheeler had given him the impression that great stress was laid by the Ordnance Department on the ratio of "non-productive" to "productive" labor at the various arsenals, as an index of efficiency, and that Wheeler for this reason had not found it possible to increase the overhead force at Watertown to the extent he considered economically desirable. This criterion of efficiency, Barth asserted, was completely erroneous. The only correct criterion was comparisons of total costs per unit of output, not comparisons of one category of costs with another. It was important, he believed, that this confusion be cleared up, for the Taylor system called for what was likely to strike the uninitiated as an astonishing number of "non-producers" and would therefore tend to raise the ratio of overhead costs to direct labor costs. But this was no reason for concern, since "under a Taylor System of Management the ratio between indirect and direct labor . . . will be absolutely incomparable with this ratio under the kind of management and the system of cost accounting now in operation at the Watertown Arsenal."

Returning to his analysis of the concrete changes which were necessary, Barth proposed to rearrange the machines on the second floor of the machine shop to clear a space of about 1500 square feet for the erection of a planning room, from which the routing of work in the shops would be controlled. The toolroom would have to be moved to a much more commodious location and a complete separation made between the tool cage (from which tools would be issued) and the toolmaking and tool-repair department, with a different man in charge of each. While these changes were being made, he would design and have printed the necessary blank forms, "of which there are a great number," so that these would be ready as soon as the planning room was completed.

Barth also proposed to make an early start with the detailed analysis of the various gun carriages then being built or on order, so that route sheets could be prepared showing the operations required on every piece of a carriage. From these route sheets would be derived the job cards, move cards, material authorizations, and so on, by which the planning room would control the flow of production. Initially the planning room should deal in detail with the machine shop only, leaving the pattern shop, the smith shop, and the foundry organized as in the past, except that they would receive their instructions on work required from the planning room instead of from the main office. Control from the planning room would be extended to these departments as soon as the routing of work in the machine shop was running smoothly.

Once the planning room was in operation and the routing of work established, Barth proposed to start with what he called the diagraming of the equipment in the machine shop: a detailed study of the characteristics and performance of each machine tool. At this stage it would be necessary to redesign and rebuild certain of the machines and to introduce procedures for the systematic maintenance of the belting. This would culminate in the preparation of a Barth slide rule for each machine, by means of which the best speed and feed for each type of work on that machine could be computed. At the same time all the appliances and tools used in the shops would be standardized, so that the functional relationships already worked out by Barth and incorporated in these slide rules could be utilized without further experimentation. To facilitate the standardization of tools and belting, Barth recommended that the arsenal should purchase from the Tabor Manufacturing Company, in which Frederick Taylor held a financial interest,[18] certain tool-grinding, forging, and belt-maintenance equipment which had been specially designed for use in conjunction with the Taylor system of management.

All this preparatory work, which Barth estimated would occupy between twelve and fifteen months, was, he asserted, the necessary preliminary to the introduction of time study, the establishment of standard work tasks, and the institution of an incentive wage system. The respeeding of machine tools, the preparation of slide rules, and the systematization of belt maintenance would make it possible to standardize and predetermine machine times for each job; but the machine time was frequently the lesser part of the total time. Task setting, "the ultimate aim and most paying part of the whole Taylor System of Management," demanded that handling times be predetermined also, and for this stop-watch time study would be necessary.

Barth concluded his report by estimating that two to three years would be required to establish the Taylor system in all the departments, while four years would elapse before the greater part of the work would be on a task basis. The cost of the installation, including the fee for his own services, would average $1400 per month for the first five months, $1100 per month for the next four months, and $600 per month during the remainder of the first year and the whole second year. After that date the expenses of maintaining and extending the system could be charged to the general overhead expenses of the arsenal. A total of $13,200 would therefore be required to finance the first year. In addition it would probably be desirable to set aside some $7000 to $10,000 for the purchase of new tools and the respeeding and rebuilding of old machines. In all, Barth estimated that it would be advisable to have between $20,000 and $25,000 available for the first year's work.

Barth's proposals, to sum up, fell into four parts or phases: (1) reorganization of the storeroom and toolroom, (2) creation of a planning room and establishment of a routing system, (3) respeeding and standardization of machine tools, and (4) installation of an incentive wage system based on time study and task setting. Barth had nothing to say in this report about possible labor trouble, beyond the stereotyped Taylorist formula that no difficulty with the workmen was to be anticipated. So many of the changes made would be to the benefit of the workman, he argued, that "by the time we get to the matters that directly affect him [*i.e.*, task setting and wages] he is usually no longer suspicious of our intentions, and therefore readily falls into line." When difficulties with employees did arise, Barth added, they were usually due to "the failure of the systematizer to go about the matter in the right way."

All this was conventional Taylor doctrine. In a letter to Crozier written shortly before Barth submitted his report,[14] Taylor also minimized any possibility of labor trouble. The basic difficulty at Watertown, he wrote, was that piecework of the ordinary type was impossible because of the absence of repetitive jobs. The arsenal was in reality an engineering rather than a manufacturing establishment and therefore called for a more elaborate system of routing and a higher order of foremanship than was needed in an ordinary manufacturing concern. "For this reason," he continued, "our system of management, if applied to the Watertown Arsenal, ought to produce better and larger results than if applied to the arsenals in which manufacturing is chiefly done." A good start had already been made by the officers in charge, and their success made it perfectly clear that there would be "practically very little trouble" with the workmen. "I should say without hesitation," concluded Taylor, "that the introduction of our System of Management in the Arsenal would result in your being able to turn out fully twice the amount of work in a given time with the same number of workmen as you now have, *i.e.*, doubling the output of the arsenal — or turning out your present output with half the men now employed."

Neither Barth's report nor Taylor's could be described as a penetrating analysis of the reasons for low productivity and high costs at the arsenal. Their analyses of the labor force, its attitudes and probable response to a cost-reduction program, were especially superficial. From their point of view any really deep-searching investigation of the current organization of the arsenal would have been superfluous. As Barth in particular implied, installation of the Taylor system meant that this organization would be superseded and a new pattern of control established. What they were offering to the Ordnance Department was not a set of proposals for piecemeal reform but a completely developed system. The results of the installation of this system were predictable, not problematic, in their view. Consequently, they felt no need to explore the existing situation in detail, particularly in its non-mechanical aspects. The careful analysis which Barth proposed to make of the arsenal's mechanical equipment contrasts vividly with his easy assumption that its human equipment would present no problems.

The officers of the arsenal and of the Ordnance Department appear to have had a broader conception both of the reasons for low productivity and of the probable complexities of introducing radical change. Colonel Wheeler, though he took the initiative in proposing the employment of a civilian expert and though he gave Barth full support when, in July 1909, he began his work at the arsenal, seems to have felt that the cooperation, or at least the passive compliance, of the employees was indispensable. In May 1909 he took the unusual step of calling together a general meeting of the employees in Watertown Town Hall, where he read to them an analysis of the problem and a summary of the measures which he proposed to take to deal with it.[15] He made no reference to soldiering, admitting frankly that the principal trouble at the arsenal was bad management. But his listeners were left in no doubt that Watertown Arsenal was regarded as the only black spot in the otherwise good record of the Ordnance Department. He quoted General Crozier's admission before Congress that "some of the work up there at Watertown has been quite expensive, and it is almost the only arsenal at which we have not always made a favorable showing in comparison with the outside establishments" and proceeded to list twenty-four reasons for this state of affairs. These were:

1. Frequent changes in management.
2. Absence of system and shop management.
3. The number of holidays and vacations with pay.
4. Lack of a proper system of supplies.
5. The conduct of work on the day wage system.
6. Restrictions imposed by laws and regulations, especially in regard to procurement of materials.
7. Lack of coordination of the work done in different shops.
8. Lack of sufficient tools of proper power.
9. Multitudinous duties of foremen.
10. Loss of time in looking for proper tools and fixtures.
11. Loss of time due to employees' waiting at grinders and at the toolroom.
12. Loss of time due to breakages or repairs of machines and belts.
13. Loss of time waiting for the next job.
14. Losses due to lack of proper instructions or to spoiled work.
15. Lack of proper toolroom equipment.

16. Lack of proper transportation and lifting equipment in the shops.
17. Wastage and lack of economy in the operation of the power plant.
18. Lack of proper attention to costs of detailed operations.
19. Endeavors to make parts with poor facilities and at great expense which could be purchased more cheaply from outside suppliers.
20. Delays in getting materials when needed and consequent changes in plans.
21. Additional cost of transport service between shops under a system which permitted one helper to each teamster.
22. Excessive amounts of metal left on castings before machining.
23. The commencement of work before a sufficient supply of materials was on hand to finish the job.
24. Failure to take full advantage of the machines and tools provided and ignorance as to the best practices.

A system which would remedy these anomalies was, he informed his audience, already under consideration. Continuation of the present inefficiencies would not only be a disgrace to the Ordnance Department but would also raise serious questions as to whether the arsenal should be closed down entirely. "If it should happen that expenditures at any establishment have not been properly safeguarded or work has not been carried out economically, Congress would certainly not only have the right, but it should be its duty, to consider whether the retention of such an establishment were worth the additional expense involved."

Carl Barth began installing the Taylor system at Watertown in June 1909.[16] Exempted by presidential order from the usual civil service rules governing the employment of civilians by the Ordnance Department, he spent during his first year an average of ten days a month at the arsenal, receiving a fee of $50 per day plus his expenses.[17] Major Williams, officer in charge of the shops at Watertown, was instructed to cooperate with Barth in every way consistent with his official responsibilities and to familiarize himself with the Taylor system so that he would be competent to maintain and extend the new procedures after Barth's departure. Barth lived with Major and Mrs. Williams while working at the arsenal and paid directly to them the money he received to cover his living expenses;[18] they appear to have become friends, and it is probable that this close personal relationship was partly responsible for the absence of friction between Barth and the officers. Colonel Wheeler, the commanding officer, was instructed to submit monthly reports to the Chief of Ordnance on the progress of Barth's work and did so until these instructions were canceled after the first year of the installation.

Barth interpreted his duties at Watertown in broad terms. He stated later:

I have looked upon my position as implying that it was up to me to follow my natural inclinations to make myself useful to the Commanding Officer of the Arsenal in every way possible. . . . In fact, it has been tacitly understood that I have been here to take the initiative in the introduction of the so-called Taylor System of Management, and to instruct both the officers and the civilians at the Arsenal in the principles and working of the system.[19]

His procedure followed closely the sequence outlined in his initial report to the Chief of Ordnance. The first point of attack was the storeroom, the inefficient management of which had been largely responsible for the arsenal's high material costs.[20] Here a new system of accounting for stores issued was instituted which will be described in detail later; its effect was to provide automatic checks against excessive or duplicate issues of material. Barth also recommended the installation of the "double bin" system,[21] whereby two separate but adjoining bins (or compartments in the case of larger articles) were used for each article in store, one the receiving and the other the issuing bin. When all the articles in the issuing bin had been distributed, it became the receiving bin and what had been the receiving bin became the issuing bin. As the bins were successively emptied, the tags showing all issues from them were sent to the property division, where they were checked against the stock sheets. This simple but highly effective system provided an automatic inventory of stores: the quantity of an article on hand was verified each time the issuing bin for that article was emptied.

An appropriation of $3345.73 for the installation of the double bin system was received in November 1909, and the change-over was begun in the following month.[22] The opportunity was taken to regroup, recount, and reclassify under mnemonic symbols all articles in stock. Progress was slow, however, as it was considered important that the regular work of the storeroom not be interrupted, and the new system was not completely in operation until September 1910.[23]

At the same time Barth began organizing his planning room — the department which was to be the nerve center of the whole organization. One preliminary difficulty arose: the engineering division, which Wheeler had only recently initiated, was already performing certain of the functions which, in the normal Taylor organizational plan, should have been centralized in the planning room (the preparation of drawings, the listing of operations and material requirements, and so on). Barth could have converted the engineering division into a planning room by extending its functions and enlarging its staff, but the engineering division was located in the administration building, whereas it was highly desirable that the planning room should be located in the machine shop. Nor could the engineering division be brought over to the machine shop, since space there was severely limited. Barth therefore decided to leave the engineering

division as it was and to create a planning room which would
be organizationally and geographically separate.

Space for the planning room was found on the second
floor of the machine shop, directly over the engine room.
Desks, boards, files, and all the other paraphernalia were
ordered to Barth's specifications, and blank forms of about
twenty different types — job cards, route sheets, storehouse
sheets, storehouse tags, routing tags, and so on — were
printed. The electric time-recording clocks, which Wheeler
had so recently congratulated himself for installing, were
discarded on the grounds that they did not fit the job cards
used by the Taylor system and that all recording of times
would be done in the planning room.[24] A new electric timing
system, consisting of a master clock, four electric time stamps,
four secondary clocks, batteries, and so on, was installed.
By the end of January 1910, the planning room was completed and ready to begin functioning.[25]

In terms of personnel, the planning room was built up
around the master mechanic's office.[26] Nelson, the master
mechanic, was relieved of his duties on the floor of the shops
and placed in charge of the planning room. North, the foreman of the machine shop, and later two other foremen were
detailed to act as his assistants. To take over North's former
duties, three workmen were promoted to the position of
"gang bosses," and each was allotted a section of the machine
shop to supervise in accordance with the instructions which
he received from the planning room.[27]

The toolroom also was reorganized along the lines laid
down by Barth in his initial report. The toolmaking section
was separated from the tool-issue section and a foreman
appointed to supervise the manufacture and care of tools.
On June 22, 1909, orders had been placed with the Tabor
Manufacturing Company for the standard Taylor tool-
forging and -grinding equipment, and this was received by

early September, although no immediate use was made of it.[28] In February 1909, the arsenal had received a special allotment of $5000 for the installation of tool-managing facilities. This was supplemented during the early months of 1910 by a further allotment for the purchase of high-speed tools to the amount of $5000 (half the minimum amount which Barth had stated to be necessary). The re-organization and partial re-equipment of the toolroom was completed by the end of April 1910.[29]

An important series of changes made during the same period was the establishment of standard procedures for the inspection and maintenance of the belting which drove the machine tools.[30] The state of the belting at the arsenal had been the subject of particularly acid criticism from Barth in his initial report. Describing it as "about the worst thing I came across there," he had particularly condemned as "antiquated" the use of rawhide lacing and had scored the absence of attempts to secure proper tension on each belt. He recommended the purchase (also from the Tabor Company) of a Gulowsen belt bench, a set of Taylor belt scales, and a Jackson wire-lacing machine.[31] This equipment had been received and put into operation by early September 1909.[32] At the same time a special belt-maintenance gang was formed, and its foreman, H. C. Knowlton, was sent for instruction to the Yale & Towne Company, of Stamford, Connecticut.[33] A great deal of the old belting was replaced with new and in some cases heavier belting. This made it possible to run the machine tools at higher speeds and with greater power, so that full advantage could be taken of the cutting powers of high-speed steel, and also prepared the way for Barth's later standardization of cutting speeds and feeds. By the end of April 1910 the belt-maintenance system was in full operation and belt failures during working hours had been practically eliminated.

In February 1909 Colonel Wheeler had written to General

Crozier to request that certain six-inch gun carriages be assigned to Watertown for manufacture, so that the efficiency of his new cost- and time-keeping system might be tested.[34] Watertown received the order, and in September 1909 it was decided that two of these gun carriages should be the first products to be put through the machine shop "under the Taylor system." [35] This meant, primarily, that their manufacture would be routed from the planning room, since Barth had not yet begun his work on the machine tools and no time studies had been made. The two six-inch carriages were to serve as pilot projects, principally to give the planning room staff and the machinists their first taste of centralized routing.

A considerable amount of preparatory work was necessary. Assembly charts were drawn up, containing a detailed analysis of the operations required to produce each component which went into a complete gun carriage. On the basis of these charts the planning room staff decided upon the sequence of operations which each component was to follow, the dates at which each operation should be started and completed, and the order in which each component should be moved from workplace to workplace, up to final assembly. These individual analyses were then brought together on a single route sheet, which formed the master timetable for the whole project. Besides the sequence of operations to be performed on each part, the planning room prescribed the materials to be used and prepared any special drawings or instructions necessary for particular operations. The route sheet was then turned over to clerks, who prepared the individual job cards, move tickets, and so on which would be required for the execution of the work. These cards, together with the master route sheet, were then passed to the route-sheet clerk, who filed them ready for use when required.[36]

This preparatory work, described here in summary form, occupied the planning room staff, with Barth's assistance, during the months of September and October 1909. There were a few initial difficulties. Nelson, the master mechanic, was accustomed to spending most of his time on the floor of the shop and was inclined to give inadequate attention to the work of the planning room, so that more than once his instructions and the formal orders from the planning room came into conflict.[37] There was a tendency to bypass the planning room, to reduce the routing system to a ritual, and to carry on the work by the old informal methods with which everybody was familiar. These tendencies were blocked, in part, by making Nelson sign all route sheets, thus insuring that he took responsibility for the routing instructions, and by insisting upon strict adherence to all orders from the planning room. And, with minor exceptions, the system appears to have worked. By the middle of April 1910 practically all the work in the machine shop was being routed from the planning room, and in June 1911 the system was extended to the smith shop.[38] The foundry and the pattern shop were still not included in the routing system as late as 1915.[39]

While the planning room was getting into action, Barth took up the part of the Taylor system which he liked best and at which he excelled: the rehabilitation of the machine tools. This involved four principal phases: the diagraming of the individual machine tools, the rebuilding and redesigning of obsolete or unsuitable tools, the standardization of ancillary equipment, and the prescribing of speeds and feeds.[40] Certain data on the performance capabilities of the machine tools had already been collected by Wheeler in the months before Barth's arrival, and more were obtained under Barth's instructions. An extensive series of tests on different types of high-speed steel was conducted and, during a slack period of work early in 1910, the opportunity was taken to relocate a number of the machine tools, to bring together in one section of the shop all equipment of the same type.

Several of the larger machine tools were provided with individual electric motors, while the smaller ones were arranged so that they could be driven in groups.[41] The electric generating capacity was increased.[42]

Diagrams were prepared for all machines in the machine shop, showing the driving arrangement, feed gears, and so on. This process necessitated the measuring of all the machines, their gears and pulleys, and the study of the diagrams to insure uniformity and the proper relationships in their speeds and feeds. A considerable amount of redesigning and rebuilding was done, particularly of cone pulleys and gears, first on the lathes and then on the drills, planers, shapers, slotters, and other machines.[43]

At the same time arrangements were made to standardize all the ancillary equipment used on the machine tools. The sockets for boring bars were standardized so that all boring bars would fit all sockets, and the slots in the faceplates of the lathes, planers, and other tools were cut to a single size. All the tee bolts were made to fit this standard slot, to avoid any delay in finding the right bolt for a particular job. To achieve uniformity in cutting tools, the workmen were forbidden to grind their own tools, as they had been accustomed to do. Instead, the toolmaking department was to forge and grind all tools to the standard Taylor specifications and by the use of Taylor-designed equipment. The tool posts on the lathes were altered and strengthened so that they could be used with these tools.

From August 1910 until his employment at the arsenal was terminated in June 1912, Barth spent four days each month at Watertown.[44] These monthly visits were entirely taken up by work on the preparation of slide rules for the machine tools and in instructing the planning room in their use. Much of the work in the machine shop was highly detailed and technical. This means that the surviving records do not provide a full account of Barth's work and that the significance of what he did is not easily appreciated by those who have little technical training.

In outline, however, the problem that Barth faced and the solution that he provided are clear enough. Apart from its organizational defects, the machine shop at Watertown was technically far from up-to-date in 1909. Into this machine shop there was introduced a major innovation, high-speed tool steel, probably the most revolutionary change in machine-shop practice within the memory of anyone living at the time. This major innovation, if it was to be effectively utilized at Watertown, made necessary a whole series of minor innovations. The Taylor system of management, which included most of these minor innovations, was essentially a means of adjusting the arsenal to the impact of high-speed steel. The order in which Barth set about his tasks should not deceive us. The reorganization of the storeroom and toolroom, the rehabilitation of the belting, and the establishment of the planning room were necessary reforms, and it was essential that they should be done first. If they had not been, and if Barth had set about respeeding the machine shop before they were done, the arsenal would have burst at the seams like an overinflated football.

High-speed steel was no minor change which could be introduced in one section of the arsenal and then forgotten. The whole arsenal had to be geared to the pace which it set. The limiting factor in productivity was no longer the machine, but the organization. It is perhaps a little difficult for the nontechnical reader to appreciate the magnitude of its effects. A description of the results of one of Colonel Wheeler's experiments may help. Wheeler took a lathe which, using one of the old-style carbon steel tools, could be made to remove a maximum of two hundred pounds of metal from a casting in an eight-hour working day. He put a high-speed tool in the lathe and set it to work on the same job under the same conditions, except that the speed, feed, and depth of cut were altered to suit the new tool. The lathe removed precisely ten times as much metal — two thousand pounds — in the same period of time.[45] This was probably an exceptional case, for the usual increase seems to have been 200 or 300 per cent.

It was possible for a machine shop to purchase a stock of high-speed tools, use them in its metal-cutting operations, and yet continue to turn out work at much the same rate as before. Probably this was what most machine shops did when the news got around of what Taylor and his friends had done at Bethlehem.[46] A shop which did this would find the results disappointing; the only obvious change would be that the tools would not need to be reground so frequently. To purchase high-speed tools was one thing; to use them correctly, so that their full potentialities could be realized, was another.

A machine shop which adopted high-speed steel and knew what the new steel could do was faced with the necessity of a complete reorganization. First, few of the machine tools built to use the old steels could be run at the pace and with the power which the new steel demanded. Hence the necessity for rebuilding and redesigning machine tools, systematizing belt maintenance and repair, and increasing power capacity. Secondly — a considerably more intractable problem — few of the machinists and foremen who had grown up in the carbon steel era had any conception of what the new steels could do. Hence the necessity for Barth's slide rules and the prescribing *by management* of speeds and feeds which, to men of the older generation, were literally fantastic. And third, since the use of high-speed steel meant very large reductions in machining times, handling times (the time taken to set up a job in the machine before machining and to remove it afterwards) came to represent a higher proportion of total job times. Hence the necessity for time and motion study and incentive-payments schemes.

But the use of high-speed steel also called for changes in organization, both formal and informal. A type of organization which would operate well enough with carbon steel might prove unworkable when called upon to handle the greatly increased work pace of high-speed steel. The change-over from one pace to the other rendered obsolete much of the experience and practical know-how of the foremen on whom older systems of management had relied. The increased formalization of authority and communications which is a noticeable characteristic of the Taylor system, the explicit attempt to measure and standardize whatever could be measured and standardized, reflected not only the faster work pace and more critical standards of high-speed steel but also the inadequacy of the old craft skills of foremen and machinists.

Barth's problem would have been simpler if it had been possible to scrap most of the old equipment in the machine shop and install up-to-date machine tools designed for use with high-speed steel. But, at the arsenal, where every proposal for new investment had to be justified at length first to the Ordnance Department and then before the jealous eyes of congressional committees, this was not feasible. It was not what Barth had been hired for, nor indeed was it a course of action which he was in the habit of recommending. To install the Taylor system was, in the minds of many businessmen, an alternative to the replacement of capital equipment; and it was one of the tenets of Taylorism that scientific management called not for replacing equipment but for getting the most out of the equipment which already existed. The Taylor system demanded more efficient and more continuous use of equipment, not its replacement; the reduction of overhead charges by minimizing the time during which machines stood idle or underutilized, not the increase of such charges by new investment. And to Carl Barth, who, for all his vaunted sympathy with the working man, reserved his most violent epithets for those who mistreated machines, this was as it should be.

By the end of 1910 the first phase of the installation of the Taylor system at Watertown had been completed. The routing of work in the machine shop was entirely controlled from the planning room; the tool- and belt-maintenance systems were in full operation; inventory control and stores-issuing methods had been reorganized and were working smoothly. To mark the end of this first phase, the Chief of Ordnance convened at Watertown in December 1910 a meeting of the commanding officers of the five large manufacturing arsenals (Watertown, Springfield, Watervliet, Frankford, and Rock Island) "to consider the business methods of the Department as applied to Watertown Arsenal, and the advisability of the employment of any of the methods at other arsenals." [47] The board of officers submitted a report which recommended the adoption at all the arsenals of those parts of the Taylor system which had been installed up to that date at Watertown. They also advised that officers and employees from the other arsenals should visit Watertown and learn how the system worked. No mention was made at this time of employing outside consultants at the other arsenals; Watertown was to be the model which the other arsenals would copy.

It was realized that the installation of the Taylor system at Watertown was not complete. In particular, no time study or bonus system had yet been attempted. Major King of Rock Island Arsenal later stated that it was not intended to apply these latter features of the system elsewhere until General Crozier had seen how they worked out at Watertown.[48] The report of this board of officers, therefore, was far from a blanket endorsement of the Taylor system; but it did give an official stamp of approval to the specific changes which had been made.

A slack period of work which Watertown experienced during the latter half of 1910 was followed by a period of intense activity during the early part of 1911, for three months of which two shifts were working in the machine shop.[49] Barth meanwhile continued with the preparation of slide rules and by the end of January had made sufficient progress to suggest that a start might be made in prescribing speeds and feeds.[50] This was the initial step in the final phase of the Taylor system: the setting of tasks. At the same time Barth recommended that a time-study expert be hired. In response to an inquiry, Taylor urged Crozier to secure the services of Dwight Merrick, a former employee at Bethlehem who had, under Taylor's guidance, specialized in stop-watch time study and task setting. Taylor warned Crozier that Merrick's services were also being sought by the Navy Department and praised his abilities highly:

By getting Merrick as a teacher you would save, I should think, a year at least in the rapidity of your time study and similar work. He is the best detail man for this work who is at all available, and also, relative to the other men, very cheap; that is, my impression is that his price is $15 a day and expenses.[51]

A presidential order was obtained which exempted Merrick from civil service rules, and on May 8, 1911, he began work at Watertown, at the salary which Taylor had specified but without expenses.[52] Unlike Barth, Merrick worked full time at the arsenal; partly because of this, partly because of his lower salary, his status was differently regarded. Barth was accepted as an independent consultant; Merrick was regarded more as an employee. This was to have important consequences.

Merrick was hired to carry out time studies, by means of a stop watch, of the various jobs in the machine shop and to teach time-study work to certain members of the planning room staff. These time studies had three chief purposes: (1) to simplify work by the elimination of superfluous motions,

(2) to set a standard time in which each job ought to be done, and (3) to provide the basis for a payments scheme which would furnish an incentive for the workmen to do the job in the standard time. Merrick's work was essentially an extension of the improvement and standardization work which Barth had begun. What Barth had done for the machines, Merrick was now to do for the men.

But, although Taylor doctrine inclined to the opposite view of the matter, Merrick's task was considerably more complicated than anything Barth had undertaken. Barth could examine a machine tool, change its gears and its belting, and reset it to run at a higher speed in complete confidence that the machine would do what he wanted it to. But Merrick had to take some account of the fact that the men would not work at the pace he prescribed unless they wished to do so. He had to face the problem of motivation.

The answer which the Taylor system provided to this problem was an incentive-payments scheme. If you wanted men to work at a certain pace, you promised them a financial reward if they did so; the problem was no more complicated than that. The plan which it was proposed to install at the arsenal was a modification of the well-known Halsey premium plan. Why this particular payments scheme was chosen — rather than, for instance, Taylor's differential piece rate — is not entirely clear. Probably the fact that it guaranteed each man at least his daily wage, whereas Taylor's did not, was the crucial factor, since daily wage rates at the arsenal were set according to civil service regulations. The plan actually adopted, however, was not Halsey's original one but was combined with the Taylorist feature of task setting by time study.

An essential feature of an effective incentive-payments scheme is simplicity. If the precise relationship between pay and output is not clear to the workman affected, the stimulus effect is lost. On this basis, the modified Halsey plan adopted at the arsenal cannot be rated very highly. Here, for instance, is how General Crozier attempted to explain it before a congressional committee:

One of the items with which we start out is the time in which the work ought to be done. Another element we start out with is that for doing the work in that time we will give the man a premium amounting to one-third — 33⅓ per cent — of his regular day's pay. Later on I will tell you how we arrive at that; but that one-third, for the present, is arbitrary. Now, we have another arbitrary rule: that for finishing the job within another time, which I am going to mention in a minute, we will pay the man for half the time he saves; and that half is arbitrary. So we have three things to start with: the time in which the work ought to be done; the 33⅓ per cent rule, and the one-half rule. Now, I will make a statement of the whole rule. After ascertaining the time in which the work ought to be done we say we will pay the man at his regular rate for one-half minute for every minute which he shall save under another time which is arrived at, so that under this one-half rule his premium shall amount to 33⅓ per cent of his regular pay, if he does the work in the time we think he ought to do it in. That is an accurate statement of the rule, and it is as concise a one as I happen to think of at the moment.[53]

Nor did Colonel Wheeler do much better when, before the same committee, he presented an example of how the premium system worked in practice:

In the case of one time study in the foundry the minimum time, taking the summation of the elementary times consumed in making a mold, was 11 minutes. The time set for the man to earn a premium was 17.1 minutes. The man was told that if he made these molds in less than 28½ minutes he would earn a premium. In other words, the 28½ minutes is 259 per cent of the 11 minutes which is the summation of the minimum times.[54]

If the reader finds such explanations of the arsenal's payments scheme a trifle complicated, he may take consolation in the fact that the arsenal employees also found it confusing. A preference for straight piecework, or better still a straight daily wage, becomes, in these circumstances, understandable. Actually neither Crozier nor Wheeler, in their explanations, did the arsenal's payments scheme justice. It did have a logic, and it was essentially simple. For example, suppose that Merrick made a time study of a certain job and, after eliminating all waste motions, summing the elementary times, and adding the usual allowances, concluded that the job could be done in six hours. This was known as the task time. To this there was added two thirds of the task time, in this case four hours. Added together, these gave what was known as the base time, or ten hours. If the workman finished the job in the base time or any greater time, he received no premium but only his ordinary day wage. If, however, he managed to do the job in less than the base time, then he received not only his day wage but also a bonus, the amount of which was one half of the value of the time he saved. For instance, if the workman in the example managed to finish the job in eight hours, cutting two hours off the base time, then he would receive, in addition to his ordinary day wage for eight hours, a bonus equal to his day wage for one hour. He would receive, for eight hours of work, wages for nine hours. If, by working at maximum efficiency, he succeeded in reducing the time taken to the task time, the theoretical minimum set by time study, he would receive eight hours' wages for six hours' work, or a bonus of 33.33 per cent. This is what the arsenal officers meant when they said that it was their intention to give every man a bonus of one third.

A payments scheme of this type was more lenient than Taylor's differential piece rate, which not only rewarded the man who reached the standard rate of production but also penalized the man who failed to do so. The arsenal plan had no direct provision for penalties; no man, no matter how slowly he worked, received less than his regular daily wage. Indirectly, however, as was brought out in the Congressional inquiry of 1911, there was some provision for penalizing the slow or uncooperative worker: any man who persistently failed to earn premiums was liable to be reclassified at a lower daily wage rate.[55] This possibility was clearly realized by the arsenal employees.

Taylor doctrine did not entirely overlook the possibility that the introduction of time study in a plant might cause trouble. The stop watch had not yet become the symbol of all that was detestable to organized labor in the Taylor system — that was to be one of the results of what Barth and Merrick were doing at Watertown — but it was already realized among the Taylor disciples that the purposes of time study could easily be "misunderstood," and certain ways of going about the matter had become standard practice. It was considered vital that no time studies should be attempted until all working conditions had been brought up to a high level of efficiency and standardized at that level.[56] There were two reasons for this. First, if conditions were not standardized, then the job itself was not standardized and could not be scientifically timed. It would serve no purpose to set a time on a job today if tomorrow the machine might be running at a different speed, or if the workman had to wait around the window of the storeroom until his material was ready, or if he had to leave his machine idle while he ground his tools. And secondly, it was believed that there would be less resistance to time study if the men being timed had grown accustomed to seeing a whole series of changes being made in their working conditions, all of which made their work easier. Connected with this "climate of change" idea was the notion that the time-study man, by the time he began making actual time studies, should have become a familiar figure in the shop and that he should have established his authority by demonstrating his superior competence in the kind of work the men were doing. It was important that he should have a certain prestige, that he should be accepted as a man who knew what he was about and could not easily be fooled.

To guard against misunderstanding and to minimize the chances of group opposition to time study, it was also standard practice among Taylor experts to approach the men individually and secure their individual consent before attempting to time their jobs.[57] In its more sophisticated form, this gambit involved appealing to the man's craft pride — "Are you a first-class man?" — and instilling in him a feeling that he was participating in a cooperative experiment. Naturally, since the time-study man was standing at his elbow and giving him all the help he could, and since he was careful, on this occasion, to make sure that the task time was not set too strictly, the workman would within a short while begin to earn substantial premiums. It was then possible — so the theory ran — to rely upon the development of a spirit of competition among the men so that they would welcome the making of time studies on their work.

Merrick began time studies in Watertown machine shop in June 1911. There was at first no trouble of any kind, or at least none that the officers could discern, except in the case of one machinist, an elderly man who had been at the arsenal for over ten years, who complained that it made him nervous to have a man with a stop watch standing over him while he worked.[58] Time studies were discontinued in this case. Barth had previously talked to the three machinists who were the first to be subjected to time study and had given them his usual explanation of what was intended, and there appears to have been no significant resistance.

Barth was by this time a well-known figure in the machine shop — he had been working there for two years — and most of his work had not only been clearly beneficial but also directly under the observation of the machinists. The new series of time studies must have seemed not very different from the studies he had already been making of machine tool capacities and speeds. Unlike other parts of the Taylor system which, as soon as they were installed, affected the organization and operation of the entire shop, time study was introduced gradually and almost imperceptibly. It was, so to speak, a matter of infiltration rather than invasion. There was no formal beginning; Barth merely had to take a machinist aside one day and say that this time he wanted to try something a little different. If protests were to be made, they had to be made as soon as the stop watch made its appearance in the shop. This demanded a considerable degree of leadership and group solidarity among the workmen.

Meanwhile a number of significant reforms were being carried out in other parts of the arsenal's organization. The first related to requisitioning and stores accounting. Under the system used before the installation of Taylor methods, when instructions were received for the manufacture of a certain article, a "fabrication order" was prepared in the office of the commanding officer, usually in the correspondence division, which was staffed by stenographers and typists who were fitted neither by training nor experience to estimate material requirements, nor indeed anything else relating to manufacturing processes. The order was in operation as soon as this fabrication order was issued. A copy of the order was sent to the foreman of each shop concerned, and he, among his many other duties, was expected to make out a detailed requisition for the material needed to execute it. The expectation was that this requisition would be accurate, as the foreman was assumed to know more about material requirements than anyone else.[59]

Actually, the requisition was usually considerably inflated, as well as being slow in arrival. It was normal for the foreman to requisition more material than he needed, to cover himself against the risk of underestimating and because it was understood in the shops that the excess material would be available for small jobs and repairs, without the formality of drawing it from stores. As the arsenal kept only a very small reserve stock of material, most of what was requisitioned had to be purchased in small quantities at relatively high prices, with a delay of from thirty to sixty days before delivery.[60]

Furthermore, it appears that it was not uncommon for requisitions to be submitted in duplicate. A foreman, having put in a requisition for material for a certain order, would be apt to submit a second requisition if the material did not arrive when he expected it or soon after he asked for it, and it was not easy to detect the second requisition as a duplicate. Material would therefore be bought twice, so that stock gradually accumulated. This was wasteful, but in addition, since the arsenal was not authorized by law to hold stocks of materials, the individual order was charged with all the material that was ordered, whether the requisition had been submitted in duplicate or not.[61]

This was a system, then, which relied entirely upon the foreman, there being no check on his judgment. What was charged against the order as "material cost" was not the material used but the material ordered. The chronic tendency to inflate and sometimes to duplicate requisitions meant that the material cost of each order was typically greater than the cost of the materials actually used in the manufacture of that order. The surplus material did not disappear; it was, theoretically, available for use on other orders. But as it was usually odds and ends and as there was no systematic check on its disposal (since it had already been "paid for"), it tended to lie around the storeroom in some forgotten corner and thus to drop out of sight from the cost-accounting point of view.

Important economies in material costs resulted from remedying the obvious defects of this system, in particular from the innovation of charging to each order only the material actually used for that order. These were not economies in utilizing materials — in the sense, for example, of manufacturing a certain article out of fifty pounds of metal instead of a hundred pounds. They were, rather, economies in requisitioning materials and in allocating their costs.

The principal changes made were two in number. First, authority was secured for the arsenal to hold inventories of materials — a change which Wheeler spoke of in 1915 as having been made "very recently." This new policy enabled material to be bought in larger quantities, with benefits in prices paid and in the speed with which material could be issued to the shops. The second change came as a consequence of the formation of the engineering division. The functions of this division were to see that all plans, patterns, blueprints, and bills of material were prepared for each order before it was issued to the foremen. This removed from the foremen the responsibility for preparing estimates of material requirements and thus diminished the risk of overbuying.[62]

Closely connected with these new methods and with Barth's work on job programing were reforms in the arsenal's system of cost accounting. This was a normal feature of the installation of Taylor methods. The Taylor system not only required prompt and accurate reporting of costs; it also generated, as a by-product, data on costs that made quicker and more accurate accounting possible. At the arsenal, reforms in cost accounting were badly needed: the difficulties Crozier experienced in securing reliable information on costs were largely due to continued reliance on obsolete cost-accounting methods. Cost accounting was used not as a method of managerial control but solely for the preparation of reports on costs of completed jobs, and indirectly of estimates for future jobs.[63] No day-to-day use was made of cost accounting to control operations.

The arsenals incurred no selling or advertising costs (except through advertising for labor), had no bonded debt, paid no taxes, and made no profits. These categories of costs were therefore irrelevant. The arsenals and the Ordnance Department did, however, have certain costs which did not vary in proportion to output or which could not be allocated directly to individual jobs or expenditure orders. Each department of each arsenal incurred costs for supervision, maintenance, repairs, purchase of machinery, heat, light, power, holidays with pay, and so on. Each arsenal incurred costs of interest on capital invested, administration, depreciation, general repairs, and the pay of officers and enlisted men of the Army engaged in production. And the Ordnance Department had to bear its share of the total expenses of the War Department in Washington. All these costs had finally, by some method or other, to be charged against the products manufactured at the government arsenals; each job that was undertaken had to bear its share of the charges, above the costs of the labor and materials which went directly into its manufacture.

Costs of production as estimated at a government arsenal (the so-called "War Department cost" was larger by the addition of other overhead costs) included three categories: (1) direct labor, (2) direct material, and (3) shop expenses and other overheads chargeable at the arsenal. Direct labor and direct materials were labor and material costs which could be attributed to a specific product or expenditure order. Shop expenses included (a) indirect or "unproductive" labor (foremen, helpers, cranemen, clerks, and similar workers whose wages could not conveniently be attributed to specific products), (b) indirect material (heat, light, power, and general supplies such as cleaning rags and lubricating oil), (c) costs of normal maintenance and repair of existing equipment, (d) costs of procuring or manufacturing new machinery, if approved by the Chief of Ordnance, (e) extra costs of manufacture arising from replacement of defective material, (f) payments to employees for legal holidays and leaves of absence with pay, and (g) costs of advertising for labor.[64]

These shop expenses had to be charged against the various articles which the arsenal manufactured so that the total overhead costs of the various departments and of the arsenal as a whole were covered. Good cost-accounting practice demanded that no one class of product should carry more than its proper share of the overhead expenses. In view of the variety of products which the arsenal manufactured, this presented the officers in charge with a somewhat complex problem in cost accounting.

Up to 1906 a government arsenal was not authorized to set up any separate category of shop expenses.[65] Under these conditions the "cost" of any job represented a hodgepodge of direct and indirect charges, there being no way of insuring that each product bore its appropriate share of the overhead costs. This was hardly a system of cost accounting; it was an absence of system. All labor and material was charged directly to orders. The result seems to have been that the larger jobs were loaded with a greater share of the overheads than should have been the case.

Beginning in 1906 Congress authorized the arsenals to set up a category of shop expenses and to charge these against individual jobs in proportion to the direct labor costs of the jobs. This was the then common percentage on wages method of distributing overheads, a simple system with a long history. It gave relatively accurate cost estimates in shops which produced a uniform product and employed a uniform grade of labor. At that time, however, it was meeting damaging criticism from professional cost accountants, particularly when applied to shops using heavy machinery, producing varied products, and employing different grades of labor with different wage rates, since these conditions made it unreasonable to use direct labor costs as a basis for apportioning overheads.[66]

Apart from its inherent defects, at Watertown Arsenal the percentage on wages method seems to have been applied in a crude and unsatisfactory manner. To quote Colonel Wheeler:

The department was . . . so wedded to this old system of charging everything directly to orders and was otherwise so conservative that for a long period after Congress authorized the taking of a ratable share of productive labor to pay shop expenses, there was a feeling that the efficiency of an establishment was measured by the smallness of the shop expense percentage and great stress was laid upon the ratio of the non-productive to the productive labor at the various arsenals.[67]

This meant in practice that as many men as possible were included in the direct labor category and their wages charged directly to orders, so that overhead charges were consistently underestimated. Furthermore, the shop expense percentages were calculated as a flat rate for the whole arsenal, instead of being differentiated by department or type of job. In the foundry, the same rate would be applied to a small bench job, such as a mold for a saddle pommel, as to a large job, such as the base ring for a disappearing gun carriage. Yet the large job called for more supervision, more power, more space, and more machinery per man-hour than did the small one. To take a hypothetical case, a job done on fully automatic machinery would have had no labor costs and therefore would have carried none of the shop expenses. Under this system, then, there was a tendency to distort the costs of different types of product: the total costs of the larger products would be underestimated and those of the smaller ones overestimated. Nor was there great improvement when, beginning in 1908–1909, separate shop expense ratios were calculated for each department.

After 1909, when Barth began his work at Watertown, a start was made with the adoption of a more sophisticated cost-accounting system. At first it applied only to the machine shop, where it emerged as an offshoot of the work of the planning department. This system was based on the use of machine rates, and had as its fundamental characteristic the computation of standard hourly charges for each machine (sometimes for each group of machines or production center). The basis for allocating overheads to a particular job was shifted from direct labor costs to the length of time each machine was utilized for that job.[68] These machine rates were intended to represent, and when carefully computed did represent, a close approximation to the actual overhead charges — including power, space, light, supervision, and so on — to which each machine gave rise in each hour of its operation. As a cost-accounting system the machine rate was particularly suitable for plants which used relatively large amounts of heavy machinery, turned out a variety of different products, and employed several different grades of labor.

This new machine rate method, as applied at Watertown Arsenal, distributed overhead charges to individual jobs in proportion to machine times rather than in proportion to direct labor costs.[69] The older percentage on wages method tended to underestimate the cost of large jobs which utilized substantial amounts of heavy machinery and to overestimate the cost of smaller jobs which utilized less machinery but more labor. The effect of the change to the machine rate method would be, other things being equal, to raise the reported costs of the larger jobs and lower the reported costs of the smaller. When comparisons were made of costs of production at Watertown before and after the introduction of the Taylor system, greater percentage reductions in costs were claimed for smaller jobs than for larger ones.

Whether one system of cost accounting is better than

another depends upon its effectiveness, relative to the cost of installing and maintaining it, as a means of facilitating economy in production. The machine-hour method of distributing overhead expenses cost Watertown Arsenal practically nothing to install and maintain, since the data required were already being reported as part of the routine of the planning room. Every time a workman turned in his job card, labor time and machine times for that job were automatically reported and filed. The data required for work programing and for the incentive-payments system served also for cost accounting. Furthermore, these data were available promptly; they could be consulted by management, if necessary, on the following day, while weekly, monthly, and yearly summaries provided a regular and reliable check on operations and served to direct attention to processes or departments that were out of line with expectations. Reforms in cost accounting at the arsenal, though by no means as conspicuous or controversial as other aspects of the Taylor system, undoubtedly contributed substantially to the tightening up of managerial control.

Carl Barth continued to visit the arsenal monthly until June 1912; he was later employed as consultant at Rock Island Arsenal. Dwight Merrick continued in regular employment at Watertown until June 1913, by which date several of the arsenal's employees had been trained to take his place as time-study experts. By the end of June 1913, then, the Taylor system at the arsenal was on a self-sustaining basis; outside consultants were no longer required. It was the arsenal's regular system, and all that remained was to extend and complete it.

Let us take a brief look, then, at this system of management as it existed at the arsenal in the years 1914–1915. The most convenient way to do this, and one which was almost universally used by writers on management at the time, is to trace the course of an order through the organization from its initial receipt to the final delivery of the products required. This is not only a simple method of presentation; it also shows management as an ongoing process of communication, which is one of the most helpful ways of visualizing an organization.

Whenever possible standard procedures under the Taylor system will be contrasted with what had been standard procedures at the arsenal before 1909. And here a word of warning must be interjected. The evidence used to piece together this picture of the system of management which Taylorism displaced is drawn almost entirely from the testimony of the officers of the Ordnance Department. These men, however honest and well-meaning, were personally committed to the Taylor system and were proud of what had been accomplished at Watertown. They therefore tended to belittle the older system and to exaggerate the contrast with the efficiency of the Taylor system. Further, the system which Taylorism displaced had been characterized by a considerable degree of informal organization, well understood by the participants but never reduced to written formulas. Most of the managing had actually been done by the master mechanic and the foremen. In part, the officers' firm belief in the efficiency of the new system may have been due to the fact that, under Taylorism, they felt that they, and not the workmen, were really managing the arsenal for the first time. To an appreciable degree one of the side effects of Taylorism was to make persons of executive and supervisory rank regard themselves as more important people, with greater control over what happened in the plant, than before. Taylorism not only involved certain new managerial functions; it also involved having them performed by new classes of people with new titles and more clearly specified responsibilities. Every person playing a supervisory role under the Taylor system of management tended, when de-

scribing his job, to exaggerate his own importance and in-
dispensability. One must not assume too readily that by
describing the formal administrative mechanisms of the
Taylor system one is also giving an adequate account of
how the system actually operated.

Under the old system,[70] when the Ordnance Department
wished to procure certain products, a request for an estimate
of probable costs and dates of completion was sent to the
manufacturing arsenals that were equipped to manufacture
items of this kind. When such a request was received by
Watertown Arsenal (the same procedure seems to have been
followed in the other arsenals also), it was passed from the
office of the commanding officer to the master mechanic and
by him to the foremen of the shops concerned. These foremen were then expected to prepare the estimates requested.
Since they kept no formal records, their cost estimates were no
more than guesses, especially when the product had not
previously been manufactured at the arsenal. Also, when
work was slack they tended to underestimate, to keep their
departments employed. Since the foreman had no ready
means of knowing the status of work already in the shops,
except for information he could carry in his head, and since
frequently his schedule of work depended upon the progress
of other departments working under the same system, estimates of probable delivery dates were extremely unreliable.
These estimates, however, without systematic checking by
any higher authority, were forwarded to the Ordnance Department in Washington and were used by the department
in deciding whether to order the products from the arsenal
or to contract with private manufacturers who had also been
asked to submit estimates. There was no regular provision
at the arsenal for checking estimates against actual performance.

Under the Taylor system, or the form of it which existed
at Watertown in the years 1914–1915, requests for estimates

were passed by the commanding officer to the engineering division.[71] This division possessed records of the costs of all products manufactured at the arsenal since the division was established, not only in aggregate terms but also in the form of detailed cost analyses of individual parts and subassemblies. Cost records were based upon the exact computation of direct labor charges by means of the job cards which the workmen received from and returned to the planning room on starting and completing work assignments. Labor charges, as computed from the job cards, were also cross-checked against the daily attendance cards. Estimates of material costs were based either upon stores records or, for products not previously manufactured at the arsenal, upon analyses of work orders, in terms of materials and operations required, which the engineering division made for every product which the arsenal manufactured or contemplated manufacturing. Overhead charges were calculated for the machine shop by the machine rate method established following Barth's analyses of machine-tool capacity, and for other departments by the percentage on wages method as revised by Colonel Wheeler, using separate expense ratios for each department. Estimates of probable dates of completion were prepared in consultation with the planning room on the basis of the work-flow charts used by the planning room for routing and by surveys of work in process and orders pending. All estimates, of costs or delivery dates, were systematically recorded and revised in the light of actual performance.

If, on the basis of the estimates submitted, the Ordnance Department decided to manufacture the products in its own arsenals, an order was sent to the commanding officer of the arsenal concerned. Under the old system, when such an order was received, a fabrication or expenditure order (the name seems to have varied) was prepared in the office of the commanding officer, usually by the correspondence division, which stated in general terms the name of the product and the number required. Copies of this order were distributed to the foremen of the departments involved, without any additional information.[72] These foremen were then expected to do everything that was necessary to get manufacture started — obtain the proper drawings, make out requisitions for material, prepare orders on the storehouse, distribute material when delivered to the correct machine tools or work-places, assign jobs to the employees by writing out job cards, see that each employee understood the work required of him and that he had the proper drawings, tools, fixtures, and so on, and decide upon the sequence in which each order should be taken up. All these duties fell to the foreman in addition to his normal functions of maintaining discipline in his shop, supervising the work, and maintaining and repairing tools, belting, and other equipment.[73] As so much of the foreman's work was clerical, he was usually allowed one or two assistants. These were generally taken from among the less skilled employees in the shop, those who could write a good, legible script being chosen when available.

A mere listing of the functions which a foreman was expected to perform under this system suggests that most of them must have been carried out inefficiently. Colonel Wheeler summed up the situation as follows: "The direct result was . . . that the foreman, instead of performing such work as he was best fitted for, by his mechanical training and experience, was confined to a desk or to an office to such an extent that work on the floor of the shops was greatly neglected, and as a rule, took care of itself." [74] Inefficiencies seem to have been marked in the two areas where coordination between departments was indispensable: requisitioning of materials and routing. The common practice of submitting inflated or duplicate requisitions for material has already been noted. The loss of working time — expensive in terms both of labor and of machine costs — involved in unsystematic distribution of material was another important factor. The following description is probably not exaggerated:

When the material was sent for it was delivered to a shop and it was generally left near the door, for the reason that the foreman had been unable to plan ahead and decide just where that work was going to be done; he did not have the time or facility for doing that; as a result, when a man finished a particular job and wanted another one, he went to the foreman who assigned him to a job, and the custom was for the machinist to either go himself or have the assistance of a helper for his material, which he expected to find somewhere in the shop. After hunting around for the material and finding it he would himself move it to his machine. He would then get the drawings to see what work was required. After that was done he would probably go after his tools. That was an expensive operation, for the reason that there was more skill involved in it than was necessary and during the operation the machine tool was idle.[75]

Similar losses of productive time occurred as a result of unplanned routing. The foreman of a shop had no means of knowing when he could expect to receive delivery of material or semifinished products; consequently, even if he had the time and the ability, he was unable to schedule in advance the orders which came to his shop. In this respect too, poor organization of the stores and purchasing department had serious consequences, especially in the period before the arsenal was permitted to hold minimum stocks of material. "We practically had no stock of material on hand and it was necessary, upon the receipt of orders, to buy most of the material after the order was received; the dates of delivery of this material were so uncertain that frequently very great delays were experienced, and the result was that, in order to keep the workmen busy, less pressing jobs were undertaken with the idea of putting them aside as soon as the material for other more pressing jobs was received." [76] Poor coordination of work flows between foundry·and machine shop, in Watertown as in many other metalworking plants, presented a thorny problem. "It used to be that if we had an order, for instance, for five 10-inch disappearing carriages the foundry would receive its orders to make the castings and the smith shop its orders to make the forgings. Now, as fast as one casting or one forging of a particular part was completed, it was often sent into the machine shop and work started on that one and completed on only the one; the other five [he meant four] would follow on at intervals of two or three weeks or two or three months, and they would be taken up individually, and all the time of preparation for doing the job would have to be wasted for those five." [77]

Let us now trace the movement of an order through Watertown Arsenal under the Taylor system.[78] If the arsenal received an order for, say, ten gun carriages, that order in its original form was immediately sent by the commanding officer to the engineering division. This division prepared what was called an expenditure order, containing a general description of the work to be done. One copy was retained in the engineering division; one copy was sent to each department of the arsenal, including the cost-keeping department; and one copy was sent to the planning room. The receipt of an expenditure order did not mean that the work was to begin; it served merely as advance notice that expenditure for certain products had been authorized.

The engineering division also prepared a bill of material for the order, containing a detailed description of all material that would be required, a list of component parts, a set of suborders which would later be distributed to the shops, and a complete set of drawings. The bill of material was sent from the engineering division to the property division, where it was examined by the balance-of-stores department. An entry was made on the bill of material showing whether or not each item was in stock and whether the quantity in stock was sufficient to fill the order. It then went to the purchasing division, which sent orders to outside dealers and suppliers for the material which the balance-of-stores department had marked as not on hand.

The bill of material was then returned to the engineering division. It now contained not only a list of items required, but also information supplied by the property division showing what material was immediately available and by the purchasing division confirming that the material not available had been ordered. Copies of the bill of material, including this information, were then sent to the departments, including the planning room. With these copies were sent sets of drawings and the suborders. In contrast to the expenditure order, which had contained merely a general description of the purposes for which expenditure had been authorized, these suborders contained instructions on the operations required on the order in each shop. When each shop had received the bill of material, the list of parts, the set of drawings, and the suborders, it had all the information necessary to begin work as soon as it was instructed to do so.

When this stage had been reached, the functions of the engineering division were at an end, with the exception of the routine revision of estimates when final costs were ascertained. Control of the order now shifted to the planning room. (In a normal Taylor organization the engineering division would have been part of the planning room.)

The planning room had received from the engineering division a copy of the expenditure order, the bill of material, the drawings, and the list of parts. Using this information, the planning room determined the operations necessary to complete each part and the sequence in which these should be carried out. At this stage, the element of time (or, more correctly, sequence) was first introduced into the analysis of the expenditure order. Having decided upon the sequence of operations, the planning room collated their decisions on a master route sheet. This was retained in the planning room, and from it were prepared various smaller cards designed for issue to the shops. These included:

1. A job card for each operation, giving the name and suborder number of the part, the code number of the machine to be used, and the location of the machine, and leaving space for the marking of the time occupied in performing the operation. This card served two purposes: when issued to the shop it instructed the workman as to what operations were to be performed on the material delivered to him, and when returned to the planning room on the completion of the operation it provided the information necessary for the calculation of labor and overhead costs.

2. A move card for each part, which accompanied the article throughout the process of manufacture from one work station to the next. When the planning room wished work to be begun on a certain part, the move card for that part was sent to the storehouse. This served as notice to issue the material required and also instructed the storehouse to deliver it to the workplace indicated on the card for the first operation. The storehouse delivered the material and returned the move card to the planning room, thus notifying the planning room that the first move had been completed. The planning room then returned the move card to the shop where the first operation was to be performed. When this operation was completed, the article was moved to the next place indicated on the card and the move card was again returned to the planning room. The planning room sent back the card, the next operation was completed, and so on. The move card enabled the planning room to control the movement of each article and, by the transference of the card back and forth from planning room to the shops, made it possible to ascertain at any moment where any article was, how long it had been there, and what operations were still required to bring it to completion. Incidentally, the move card provided a check on inspections. When inspection was required at the conclu-

sion of any stage of manufacture, the move card had to be returned to the planning room by the inspector, with his initialed approval, rather than by the shop foreman.

3. Move tags. These were small labels attached to the article being manufactured and containing instructions to the "move men," the unskilled laborers who were responsible for the actual physical transfer of materials and products. There was a separate move tag for each move, whereas the move card listed the entire sequence.

4. Instruction cards. In other plants these were a conspicuous feature of the Taylor system, designed to convey detailed instructions to the workmen on how their job assignments should be carried out. At Watertown they were little used, chiefly because no attempt was made to dilute the arsenal's skilled labor force by the employment of semiskilled labor to perform fractions of skilled jobs. The suborders and the job cards contained general instructions for each job; these were supplemented if necessary by verbal instructions from the shop foremen or gang bosses, who were expected to exercise close supervision over the work now that many of their clerical functions had been taken over by the engineering division and the planning room.

These various sheets and cards were kept on file in the planning room until the purchasing division reported that all material ordered had been delivered and was available in the stores. It was not possible for work to be started until this notification had been received. When the report from the purchasing division arrived, or as soon after as was convenient, the planning room sent out the job cards and move tickets to each of the shops concerned. A stores-issue ticket was also sent to the storeroom, ordering the material to be delivered to the first machine involved. As soon as this material was delivered, the first job card was posted in the shop, to be undertaken as soon as the appropriate machine was free. No job cards or move tickets were issued to the foundry, since that department was not under the jurisdiction of the planning room. Relations between planning room and foundry were confined to the issue of the expenditure order, the suborder, and, when the castings were ready and delivered to the storeroom, a move card stating to what part of the machine shop they should be transferred. The routing of work began only when the castings emerged from the foundry.

Each suborder carried a number. All material was purchased under that number, all labor was performed under it, and at the end of the work all costs were computed with reference to it. Each job card carried the suborder number, as well as the code numbers of the machine and the workman assigned to the job. When the men began work on a certain article, the time was stamped on the job card, and similarly when the card was returned at the end of the job. If work on a suborder was halted or laid aside, the relevant job card was called in and stamped and a new job card issued. Cards were also called in and new ones issued at the end of each pay period.

To compute total direct labor costs on an article, it was necessary only to add the elapsed times as stamped on all the job cards which carried that suborder number and multiply by the hourly wage rates of the men concerned. To compute a man's total earnings over a given period, it was necessary only to add the times stamped on all the job cards which carried his code number, multiply by his wage rate, and add any premiums earned. Thus the same cards were used for cost keeping that were used for payroll calculations; they were merely differently shuffled. The total times stamped on all job cards issued in a given shop were checked at the end of each pay period against the total times computed from the attendance cards by which the men clocked in when arriving and leaving work.

Since the job cards carried the code numbers of the machines used on each job, it was an easy matter to compute overhead costs, using the machine rate method, by multiplying the time spent on each machine as shown on the job card by the relevant machine rate and adding the totals. The total overhead costs computed for every job were checked against the total overhead charges for the shop computed from the general financial accounts of the arsenal. Material costs were computed by totaling the costs of the items on the bill of materials for a given suborder. This figure was checked against the total costs of the items actually withdrawn from stores under that suborder number.

Thus each category of costs was checked at least once against an independent calculation. These cross-checks not only tested the accuracy of calculation; they also provided indexes of waste and underutilization of equipment. The checking of job cards against attendance cards highlighted the ratio of idle to productive time, since idle time would be included in the attendance card record but would not be charged against any suborder number. Discrepancies between overhead costs computed from job cards and total overheads computed from the financial accounts provided an index of the degree to which equipment was underutilized, since idle machinery would make no contribution to overhead costs as computed from job cards. Discrepancies between material costs computed from the bill of materials and from actual issues from stores indicated wastage or loss of material. Further, the cost records compiled by the engineering division enabled ready comparisons to be made between current, estimated, and past costs, not only for completed products but also for individual parts and operations.

The issue of material from stores was controlled by a simple but effective procedure. Duplication and inflation of requisitions were obviated by having the bill of material prepared by the engineering division and the requisition by the balance-of-stores department, instead of leaving both functions to the foremen, as before. Overdrawing of material for a particular order was guarded against by setting up, for each suborder, a special material account.[79] This was analogous to the procedure followed in banking when a customer arranges a loan, with the exception that the account was in physical rather than monetary terms. The size and content of the material account were determined from the bill of material. Whenever a shop foreman or a workman withdrew material from stores under a suborder number, the material account for that suborder was debited with the items withdrawn. The items were checked off against the material on the suborder and also against the bill of material. Thus it was impossible, without special permission, to withdraw from stores more material for an order than had been authorized.

The nerve center of this network of communication and control was the planning room. The Watertown planning room contained in March 1915 twenty men, drawn mostly from former shop foremen and skilled workers.[80] The principal persons and their duties were as follows:[81]

1. The master mechanic, who had charge of the entire planning room and also, under Major Williams, of all the work in the shops. His principal task was the preparation of assembly diagrams and route sheets, which required considerable experience, technical ability, and familiarity with the capabilities of the various shops. He had the assistance of several former foremen.

2. The route-sheet clerk, who supervised the preparation, filing, and issue of the move cards, stores-issue tickets, and job cards. It was also his duty to insure that the foremen of the shops were kept fully informed of the progress of work, so that they could verify the delivery of material at the machine and make arrangements for any special tools or fixtures required before starting a job. This he did by seeing that job cards were posted promptly in their proper order on the "gang boss boards" in the shops. He was also responsible for recording on

the route sheets when a job was posted, completed, inspected, and moved.

3. The production clerk, whose duty it was to see that the work was undertaken in the order best calculated to meet promised delivery dates. He had supervision over the procuring of material from the storeroom and over the sequence in which the various orders should be undertaken. Thus while the route-sheet clerk decided upon the sequence of operations involved in each order, the production clerk decided upon the sequence of the orders. In modern jargon he might be called an expediter; he was responsible for locating and remedying sources of delay and for determining priorities.

4. The rate-setting department, consisting of three time-study men and three rate setters. All six men were trained in time-study work, but the three rate setters were fully occupied in setting tasks, calculating premiums, and preparing instruction cards. The time-study men, of course, were responsible for conducting time studies in the shops, both for purposes of job analysis and to provide data for rate setting.

5. The "window man." This minor functionary was responsible for issuing and receiving job cards and stamping times of issue and receipt on them. He also issued the drawings for each job.

The planning room included several typists and junior clerks and had supervision over the staff of move men and messengers who distributed material and instructions. The cost-keeping department was located in the planning room.

There were only two formal functions previously performed by foremen which, under the Taylor system, were not transferred to the planning room, the engineering division, or the special belt- and machine-maintenance gangs. These were the immediate supervision of work in the shops and the maintenance of discipline, which were now performed by persons referred to as foremen or gang bosses; the former term continued to be used by the workmen, while the latter was used by the officers when they remembered to do so. These gang bosses were men who had been promoted from the ranks of the skilled workmen when the foremen were transferred to the planning room. They had fewer men to supervise than the foremen and considerably fewer functions.[82] The most important were checking up on delivery of material, obtaining special tools and fixtures, and insuring that the men understood and adhered to the instructions they received.

The transference of the foremen of the machine shop to the planning room — in fact, the complete elimination of the position of foreman in the machine shop as it had formerly been understood — meant an important change in the social structure of the shop. The gang bosses did not occupy the same position vis-à-vis the workmen as the foremen had; they had fewer functions, less authority, and far more orders to obey. Their status was considerably closer to that of the workmen. The foreman of the foundry, on the other hand, was not transferred to the planning room; his sympathies before and during the strike of 1911 seem to have been entirely with the workmen.[83]

Since the new system did not always work as it should, particularly in the early days, the gang bosses sometimes found themselves stepping outside their formal role to unsnarl difficulties with the planning room and perform functions which were formally outside their jurisdiction. This even extended to routing. The gang boss of the south wing of the machine shop stated in 1911, for example, that the routing system "has not worked out as it should have worked out, and I still have many duties to perform; in fact, one half of my duties is to hunt work for the men." [84] This is one of several items of evidence which suggest that the way things were done at Watertown under the Taylor system did not always coincide with the mechanism described. To say that the workers and foremen at Watertown carried on in much the same way under the Taylor system as they had before would be an exaggeration; some real and concrete changes resulted from the introduction of the system. But it would be equally far from the truth to assert that a description of the way the Taylor system was supposed to work — a description based largely upon the testimony of the officers and Carl Barth — tells all about what actually went on.

Chapter ***4***

CONFLICTS

Taylor and Barth interpreted their responsibility as that of introducing certain technological and administrative changes at Watertown Arsenal. In fact they were doing much more than this: they were disrupting an established social system and trying to build a new one. Nothing they did was, in this respect, neutral; nothing was merely technological or administrative.

The success achieved by the introduction of the Taylor system at Watertown depended only partly upon the technical competence of Taylor and Barth. It depended also upon the ways in which other people reacted and the extent to which these reactions were predicted, controlled, and allowed for. Many of these reactions came to a focus in connection with the molders' strike of 1911 and the controversies that followed it. In itself a small thing, the strike was the symbolic issue around which attitudes became crystallized. Conflicts which were hidden before the strike were made overt after it. Analysis of the strike provides insight into the whole range of reactions to the introduction of the Taylor system at Watertown — reactions which, if the strike had not occurred, might have long remained hidden and implicit.

Dwight Merrick, the time-study expert, arrived at Watertown Arsenal at the end of May 1911. Under Carl Barth's supervision he immediately began conducting time studies of jobs in the machine shop, and by early in August ten men in that department were working on premium jobs — that is, jobs that had been timed and put under the incentive-payments system. No protests were encountered, except from one machinist who complained that it made him nervous to have a man stand over him with a stop watch.[1]

In the meantime the officers at the arsenal, in particular Colonel Wheeler and Major Williams, were growing concerned over what they felt to be an undesirable delay in extending the Taylor system to departments other than the machine shop. It was obviously working well in the machine shop; why should it not be applied throughout all departments? In particular, could not something be done about the foundry? This was the shop that had been regarded as most substandard in productivity. Yet Barth had done no work in the foundry and had not even brought it under the control of the planning room, professing to be currently fully occupied in the machine shop. Merrick too had plenty to keep him busy making time studies on the machinists. Wheeler and Williams resolved, therefore, to see whether they could on their own initiative introduce into the foundry reforms analogous to those that Barth had been applying in the machine shop. Proceeding warily, Wheeler sent off a series of letters to men whom he knew to be experts on foundry practice or whose writings on the subject had caught his attention. Since the literature of the Taylor system included little on foundry management beyond the most general principles, perhaps other authorities could be of assistance. Letters were sent to Robert H. Rice of General Electric's turbine department at Lynn, Massachusetts; to Dr. Bradley Stoughton of New York, a steel foundry expert; and to the editor of *Iron Age*, which had recently published an article on systematic foundry operation.[2]

While this was going on, Frederick Taylor had personally taken a hand in the progress of events at Watertown. He too was anxious that no time should be lost in extending the coverage of scientific management, now that such a good start had been made. To be sure, early in June 1911 he was still advising caution. "I feel quite sure that if you go right straight ahead in introducing our system, one step after another, and do not attempt short cuts, and do not try to hurry it too fast, that you will meet with practically no opposition." [3] But later in the same month he went to Washington to see Crozier, and during that visit he expressed himself in rather different terms, suggesting that the officers at Watertown should "give some of the workmen a sort of advance benefit of a premium system by making a rough guess at the time of some of the tasks and giving them a part of the time which they may be able to save from the rough estimate." [4]

Here was a remarkable suggestion, particularly coming from Taylor. Who had been more vehement than he in condemning illegitimate deviations from the strict prescriptions of his system? Who had castigated more violently the unprincipled quacks who were willing to take short cuts, to guess when they should have measured, to use the name and dignity of science to cloak deception and ignorance? Crozier was shocked, for he was inclined to take scientific management seriously. Writing to Taylor three days later, he disagreed with the suggestion. It would be "confessedly inaccurate" and would "require special efforts to avoid the raising of false hopes." The whole idea seemed to him questionable, unless Barth and Merrick had made themselves very unpopular or unless Taylor thought that it was necessary to "put the workmen in a good frame of mind" before proceeding with time study. "The thing seems to be going on all right now, and continuing to follow the present course we apparently would not do anything which would not be pretty nearly right. I cannot see the advantage of a placatory method." [5]

Clearly Crozier did not understand the complexity of the situation as the subtle Taylor did. The essential point would have to be spelled out for him. "I fully appreciate," wrote Taylor in reply, "the undesirability of using guess methods as a regular thing, and I certainly would not think of starting the premium plan until after the men recognized clearly the nature of our regular bonus system." [6] But the situation at Watertown was unique and involved special considerations. There was, after all, an election in the following year; the unions' campaign against Taylorism was not dead yet; and it might be of some importance to consider the political aspects of the problem. The objective he had had in mind, wrote Taylor, was "to get so far along with the whole system before the critical period comes during election time next year, that the workmen at the Watertown plant would be with us heartily instead of against us, and you can do nothing which will bring them more over our way than enabling them to earn higher wages under our system than they were earning before." [7]

By making this suggestion — so strangely at variance with the principles of his own teaching — Taylor involved himself in responsibility for the trouble that was to follow. Probably the officers at Watertown did not require much urging to fall in with what he proposed; but that they would have taken this short cut on their own initiative, in view of Crozier's serious misgivings, seems doubtful. More than that, Taylor encouraged the officers to believe that deviations could with impunity be made from the strict prescriptions of his system. The lesson cannot have been lost on Wheeler and Williams as they planned the reorganization of the Watertown foundry. If any doubts remained, they were laid to rest by a conversation with Carl Barth, who told them that they could certainly go ahead with their incentive-payments scheme in the foundry if they wanted to, provided it was understood that it was no part of the Taylor system. [8]

A few years earlier, when the first steps were taken to install Taylor methods in the machine shop, the molders employed at the arsenal had talked the matter over and had informed Wheeler, through a committee, that they disapproved of such methods being adopted in the foundry. [9] They had received an answer which they interpreted to mean that Wheeler was satisfied with the output of the foundry. About the same time a statement by General Crozier had been circulated which they took to mean that he would introduce no part of the Taylor system that might be objectionable to the men. Both Wheeler and Crozier subsequently denied that they had made such statements. Wheeler claimed that he had said he was satisfied with "the progress of improvement" in the foundry and "the quality of the castings turned out"; but he had not expressed himself as satisfied with the output per man per day. Crozier claimed that his statement had been that, while he would be "guided by the interests of the Government," he did not believe "that these interests would involve anything antagonistic to the interests of its employees." [10] Clearly neither Wheeler nor Crozier had meant to say what the molders thought they had said. The molders believed, however, that they had been given assurances, first, that the productivity of the foundry was satisfactory, and second, that they would not be required to work under the Taylor system if they objected to it. By the summer of 1911 they had been left undisturbed by the Taylor system for two years; nothing gave them reason to doubt that their interpretation of what Wheeler and Crozier had said was correct.

As the arsenal's management saw it, the situation was different. The foundry, to Wheeler and Williams, was the department of which they had least reason to be proud. The machine shop, before Barth arrived, labored under the handicap of obsolete and underpowered machine tools and was not organized to take advantage of the new high-speed steels. Reasons and excuses could be found for low productivity in that department for which the machinists could hardly be held responsible. But the foundry, as they saw it, was a different matter. There were problems of overcrowding, bad light, and inadequate moving equipment; but the fundamental responsibility for low productivity lay with the molders. When Wheeler first mentioned to Crozier the possibility of adopting the Taylor system, he went to the foundry for his most flagrant example of "lack of shop management." [11] Nor had the situation improved since then. Costs of production in the foundry were consistently higher than the costs of outside contractors. Testimony given by Colonel Wheeler after the strike illustrates the assumptions with which he approached the problem. "We know and have known for some time," he said, "that our molders have not turned out as much in a day as other molders; and we have used every endeavor to increase the product of the foundry, and I must say, unsuccessfully." [12] The output of the foundry was, in his opinion, "not much more than one-half what it should be," [13] and the responsibility lay primarily with the men themselves. "In no trade that has come under my observation," he asserted, "are there so many opportunities for doing useless and unnecessary work as in that of the molders, and which are more difficult of detection and observation." [14] The fact that so little machinery was used aggravated the problem: in the machine shop the machine tools set the pace, but in the foundry the molders did. The men were, in fact, soldiering, and something had to be done to disrupt the practice.

Wheeler and Williams were by this time thoroughly familiar with the Taylor system. They had read Taylor's articles, they had watched Barth in action, and they had had ample opportunity to discuss their problems with the "experts." They cannot have been unaware of the fact that Taylor regarded piecework and other incentive-payments schemes as in themselves totally inadequate solutions to the problem of soldiering. They must have realized the significance of the long and detailed preparatory work that Barth had carried out in the machine shop before he had dared introduce the stop watch. Yet when Wheeler and Williams turned their reforming eyes on the foundry, that educational process might as well never have taken place.

The plan was to put as many of the molders on an incentive-payments plan as quickly as possible. Since there had been no time studies in the foundry, and apparently could not be any in the near future, the straight Halsey premium plan was to be used, with the standard time set by reference to past performance, not to stop-watch time study. This procedure was regarded as temporary; it was to be discarded as soon as Merrick was able to leave the machine shop and turn his attention to the foundry. There was at this time in the foundry an order for castings for twelve six-inch disappearing gun carriages. Cost records for previous orders, with the work done on a straight daily rate, were available, and no further orders were expected. It would be easy to set standard times for the castings, and the aftereffects of error would not be too serious. The data might not be scientific and might enable some of the men, for a time, to earn inordinately high wages; but possibly the men would be encouraged to break away from their customary work pace and eliminate some of the waste. As Wheeler expressed it: "I was convinced that it was to the advantage of the department to have this work done under the premium system and I was equally impressed with the fact that it would be to the advantage of the workman if he were allowed to share in the profit. And for that reason we decided to put this on the premium basis, giving the man every advantage possible on a time which we knew was longer than it probably would have been had we made a time study." [15]

Although Wheeler and Williams were clear in their own minds that they were not installing the Taylor system in the foundry, they did attempt in some degree to standardize and regularize working conditions before introducing their premium system. Efforts were made to relieve the molders of as much work that was not strictly molding as possible. A helper was detailed to look after patterns, another to have charge of flasks and keep them ready for the molders when they wanted them, another to run the sand grinder, and another to move the sand to the molders' boxes. Machines were installed to straighten "gaggers" and cut them into appropriate lengths, and arrangements were made to have a supply of gaggers of convenient sizes available at all times. An attempt was made to keep the foundry floor clear of obstructions, although the lack of moving equipment presented difficulties.[16] These improvements merely scratched the surface of the problem, and most of the work in the foundry went on in the old pattern of comfortable disorganization. As Major Williams expressed it, with diplomatic understatement, "At best we must be satisfied with more or less of a compromise. It is realized perfectly well that on casting days there is necessarily a considerable amount of confusion in the shop." [17]

On July 27, 1911, the first move was made to install the premium system. Base times were calculated, from cost records, for the gun-carriage castings, and the molders were informed that they would receive as a premium over their regular daily wage one half of the time saved from that previously required for each casting. They were assured that care would be taken to provide them with all the material required and that, wherever possible, each of them would be given a helper. Between July 27 and August 9 five of the eighteen hand molders in the foundry (there were no machine molders) — Hicklin, Goostray, Wilson, Lawson, and Fraser — worked on premium jobs.[18] Each of these men earned a premium by completing his job in less than the time set. In all cases, however, the officers considered the results extremely disappointing. As Colonel Wheeler expressed it, "It was plainly evident that but little effort was being made to turn out the molds with any material saving in time. In the case of one molder it was very apparent that he was holding back. In all of these cases, considering the extra effort on the part of the management to facilitate molding, it is plainly evident that these men were, and still are, 'soldiering.' " [19]

The situation was not without its paradoxical aspects. Before introducing the premium system, the officers had believed that the men were soldiering. Now that the molders were actually earning premiums, belief hardened into conviction. In a sense Wheeler and Williams were the victims of their own tactics. They had set times which they knew to be overgenerous; now they resented the fact that the men, without stepping up their work pace, were receiving the benefit of this generosity.

When an obstacle appears to resist pressure directed toward its removal, two alternative courses of action are open: to desist and try some other method, or to increase the pressure. Wheeler and Williams chose the latter. The initial incentive-payments scheme had not been based on scientific time study. Only by time study could deliberate wasting of time be detected and a truly effective payments system introduced. Thus the officers, who earlier had believed that they could dispense with time study, found themselves forced back to Taylor techniques. Not to the unadulterated Taylor formula, however, for working conditions in the foundry had not yet been systematized. Strict adherence to Taylor's precepts would have prohibited the introduction of the stop watch in the foundry at this juncture. If Wheeler and Williams knew this, they found it convenient to overlook it.

Dwight Merrick was at this time employed full time at the arsenal. By his own admission he knew nothing about foundries or the molder's trade; his experience had been in machine shops.[20] For the previous six years he had been working on time study and rate setting in machine shops, and it was by virtue of this experience that Taylor had recommended him for employment by the Ordnance Department. Earlier, when Wheeler and Williams had first proposed to reorganize the work of the foundry, it had been agreed that Merrick was fully occupied in the machine shop and should not be asked to transfer his attention to the foundry. Carl Barth would never have agreed to sanction time studies in the foundry, since no standardization work had yet been done there. By the first week in August 1911 the situation had changed. Barth was not there to be consulted, since his regular visit was scheduled for later in the month. Wheeler and Williams were convinced that nothing but stop-watch time study would shake the molders out of their routine. And Merrick, as a full-time paid employee, was not inclined to oppose his personal scruples to the explicit orders of his commanding officer. When on August 9 Wheeler gave instructions that time study was to begin in the foundry on the following morning, Dwight Merrick obeyed.

There were no danger signals visible. The molders showed no obvious resentment at the experiments that were being conducted on them. Larkin, foreman of the foundry, reported no undercurrent of hostility. As a consequence, special precautions were not taken. On the morning of August 10 Merrick simply appeared in the foundry and began making observations with his stop watch. The molders were given no explanation. As events were to show, they already understood what was going on, or thought they did.

Merrick selected for his first time study not one of the large gun-carriage castings that had earlier been put under the premium system but a small and relatively simple bench job: a mold for the pommel of a packsaddle, a considerable number of which were on order. Under the ordinary day wage system, these molds had taken 53 minutes each to manufacture. Merrick timed the job several times, eliminated what he believed to be the time taken by superfluous or wasteful movements, and concluded that the task time for the job was 24 minutes. In accordance with the usual two-thirds formula, this gave as the base time, at which premiums would begin, 40 minutes (24 plus two thirds of 24). The molder, a man named Hendry, was told that for each minute he saved under

40 minutes he would receive, in addition to his regular daily wage, a premium equal to his daily wage for half a minute. If he reduced the time taken to 24 minutes, he would receive a premium of one third over his daily wage.[21]

The foundry at Watertown Arsenal was a relatively small shop. Anyone present could see anything that was happening. Merrick was a stranger in the foundry, and the stop watch an instrument that, by reputation at least, was recognized and feared. If Merrick thought that his time study was unobtrusive and unobserved, he was mistaken. Every molder in the foundry was watching, openly or secretly. More than that, other time studies were being made besides Merrick's. A molder named Perkins, working nearby, was timing the same job that Merrick was. Perkins was using an ordinary pocket watch; he was unfamiliar with the sophisticated techniques by which Merrick discriminated between productive and unproductive movements, timing only the former; he was an ordinary molder who could tell the time of day and who knew that, if you noted the time when a job began and the time when it ended, you could tell how long the job had taken. So Perkins surreptitiously timed Hendry making molds for packsaddles; and Dwight Merrick did the same openly. Merrick said that the job should be done in 24 minutes; Perkins' watch said that Hendry had taken 50 minutes for the first mold and 49 for the second, and Perkins' fellow molders believed that his watch had not lied.

Merrick was an expert in time study. He knew that to obtain an accurate task time it was necessary to eliminate waste motions from the measurement. But, expert though he might be, Merrick had his limitations; he was not an experienced molder. By what right, then, did he claim to be able to distinguish between the parts of a job that were necessary and the parts that were not — to decide that the man was being overconscientious, or was using too many nails to reinforce his molds, or was washing them with unnecessary thoroughness? Even for an experienced molder, matters of this kind involved careful judgment, and opinions might legitimately differ. Yet Merrick went confidently ahead, eliminated what he thought should be eliminated, and set in the end, with every semblance of accuracy, 24 minutes as the time in which the job ought to be done.

But he was not to escape unscathed. The discrepancy between the new task time and the time it had formerly taken to do the job was a large one. Apart from the surreptitious time study made by Perkins, there was reason enough to protest. And protest Hendry did, with some vigor. The time given him, he said, was too short. He could not and would not work at that pace. Here was the first direct challenge to Merrick's authority. If time study was really scientific and accurate, the time set was not a matter for bargaining. Once decided upon, it could not be tampered with, for to do so would be to admit that the claim to objective accuracy was false. Was time study absolute or was it not? A great deal turned upon the answer.

Hendry appealed to the foreman, Larkin. Larkin appealed to the officer in charge of the shop, Major Williams, and informed him that in his opinion the time set was unreasonably low. Williams, realizing Merrick's lack of experience and impressed by the foreman's strongly expressed opinion, asked him what, in his view, would be a reasonable time. Larkin stated that the job had in the past taken about 53 minutes and that he felt 50 minutes would be fair. Williams then arbitrarily raised the base time (at which premiums would begin) from 40 minutes to 50 minutes, on the theory, as he later explained, that "if his time was increased to 50 minutes it would be entirely in his favor and put him on about the same basis as all of the other molders who were working on premium jobs in the foundry." [22]

This was a critical moment, on which more depended than anyone realized at the time. If Merrick protested against Williams' decision, there is no record of it. He had grounds for objection. A vital principle was at stake — a principle fundamental to the entire Taylor system. He had been called into the foundry because a looser type of premium system based on historic times had proved ineffective. Now his first time study was being tampered with on the nonsensical grounds that it was too stringent and that the man under observation should be put "on about the same basis" as other molders whose jobs had not been timed. More seriously, his status and authority in the foundry had been undercut by a superior officer. If Williams was prepared to manipulate the results of time study in this arbitrary fashion, what was the point in going any further? What respect would the molders give time study in future, when one of their own officers had made it clear that he did not respect it?

Merrick may have felt that it was not up to him to oppose or object to the decisions of higher authority. The occupation he practiced had not yet successfully asserted its claim to professional status, and there was no recognized code of professional ethics on which he could rely. But the fact of the matter is that Merrick himself had no faith in his own time study. His later testimony is illuminating. Questioned as to how he decided upon 24 minutes as the task time, he replied: "I was more or less at a loss to know what time to set upon the job, but I am very confident that I didn't get a true observation. . . . I felt that 30 minutes was too long a time for it, and I made a guess at 24 minutes." [28] This was scientific time study in action. A man as uncertain in his own mind as Dwight Merrick was, as conscious of his own limited knowledge, and as insecure in his claims to authority, was not likely to stand firm when his superior officer intervened.

What had the molders learned from this first exposure to time study? At the very least, that the results of time study were not inviolable. A job time could be changed, if enough fuss were made about it. No one would be able to convince them that, under the Taylor system, they were subject only to the discipline of impersonal scientific law. They had learned, too, that the time-study man was an employee like themselves, that his word was not final, and that they could, in favorable circumstances, have his verdict overruled. And above all they had learned, or thought they had, that time study was dishonest. For had not Perkins' watch told them that the job had taken 50 minutes, while Merrick said it had taken only 24?

This last lesson was the one that rankled. Authority was a familiar thing, and they could accept it; but dishonesty was something different, something unpredictable. They had always been treated fairly by the officers, and they regarded this as their due. But now, under this so-called Taylor system, a man had been brought into the shop clothed in authority, as they saw it, to determine how fast they should work and how much they should be paid. This man had no claim to their respect that they could discern, for he knew nothing about the molder's trade and had done nothing else to demonstrate his competence. His word was not to be trusted; they had caught him in what they took to be a deliberate falsification — and one gravely prejudicial to their interests. Against this sort of thing a stand had to be taken immediately, otherwise nothing could be counted on in the future. To submit without protest to this kind of treatment was not right.

Nothing further of any significance happened in the foundry that day, as far as Merrick and Williams could tell. That evening, however, after supper, the molders met privately to talk over what had happened and decide upon what they were to do if another attempt was made to take time studies on the following day. This was not a meeting of the union, although the men present were union members. It was not a meeting of any formal organization. It was unusual, for the molders were not in the habit of meeting outside working hours. They lived in different parts of Boston and its suburbs, traveling to work by streetcar every day, and they would not ordinarily see each other from the end of one working day to the start of the next. The fact that they met on this occasion reflects their sense of crisis. Nobody called the meeting; it just seemed to all of them the best thing to do.[24]

Knowledge of what took place at this meeting rests upon the later testimony of one of the molders, Joseph R. Cooney, who appears to have assumed the role of spokesman for the group. The principal topic of discussion was the time study that had been made on Hendry, and the marked discrepancy between Merrick's time for the job and the time estimated by Perkins. This was interpreted simply as dishonesty on Merrick's part. It was agreed, however, that such dishonesty was not a matter of the individual who happened to be making the time studies, but rather concerned the whole tendency of the Taylor system. If permitted to pass this time, it would occur again, as long as time study continued. Objection was made not against Merrick as a person but against time study and the premium system in general. In Cooney's words, the molders agreed that "if Mr. Taylor's methods were to be carried out in their entirety, as they are recorded in his book called Taylor's 'Scientific System of Shop Management,' they felt that they could not submit to these methods any longer." [25]

No mention was made of the possibility of going on strike. They agreed to draw up a petition, which all the molders would sign and which would be presented to the commanding officer if any further attempt were made to take time studies; and they agreed that the next man ordered to work under the stop watch should refuse to do so. The petition was drafted that night and signed by all of the molders and some of the helpers before eight o'clock the following morning. It read as follows:

Lieut. Col. C. B. Wheeler
Commanding Office of the Watertown Arsenal

Dear Sir:

The very unsatisfactory conditions which have prevailed in the
foundry among the molders for the past week or more reached an
acute stage this afternoon when a man was seen to use a stop
watch on one of the molders. This we believe to be the limit of
our endurance. It is humiliating to us, who have always tried to
give to the Government the best that was in us. This method is
un-American in principle, and we most respectfully request that
you have it discontinued at once.

We feel justified in making this request, on the ground that
some two years ago you told a committee of the molders that you
were well satisfied with the output of that department; also Gen.
Crozier gave his word to a committee that waited upon him in
Washington that he would not install any part of the Taylor system that might be objectionable to the men; and we assure you
that this part of the system will not be tolerated by the molders.

Very respectfully,

Joseph Hicklin
Isaac Goostray
Martin Roach
John Hendry
G. E. Lawson
John Weir
J. J. Flynn
J. F. Murphy
E. L. Sherman
L. Katz
Thomas Kane
A. P. Doherty
B. Hall
John Wilson
James T. Fraser
A. F. Perkins
E. A. Joyce
J. R. Cooney
John T. Sullivan
George V. Miller
John F. Gatte

With this petition ready for presentation, the molders went to work as usual on the morning of August 11. Suspecting nothing, Dwight Merrick appeared in the foundry with his stop watch at an early hour and selected the job that he intended to time that day. The molder concerned was Joseph Cooney. Cooney, with the eyes of his fellow molders on him, flatly refused to cooperate. Here is the story in his own words:

That morning the watch was put on me, and, of course, according to the agreement we came to, I told Mr. Merrick he could not take any time record on me, and that we had come to the conclusion that we would not submit to the method of taking time, so he asked me what was the matter. I said, "Nothing in particular, only that we don't care to stand for this." "Now," he says, "my good fellow, go right along just as though I wasn't here at all"; and I said, "Either you or I will have to be on the floor alone." He said, "I don't care to enter into any argument with you," and I told him that I did not care to enter into any with him, and that there was no necessity for it. I told him that we had a statement prepared, and if he wished to take it to the commanding officer he could do so, or if not we would give it to the foreman.[26]

Fortunately for Merrick, the matter was not left to his discretion. While the argument was going on, another of the molders, Isaac Goostray, took out the petition and handed it to the foreman, Larkin. Larkin gave it to Major Williams, who immediately left the foundry and went off in the direction of the main office to find Colonel Wheeler. Wheeler told Williams to explain the situation to Cooney and to inform him that as long as he was employed at the arsenal he was to obey orders. After about fifteen minutes, during which Cooney continued to work at his job, Williams reappeared in the foundry and summoned Cooney to his office. Cooney's narrative continues:

I was called into the foundry office and Maj. Williams explained to me the reason why time record was being taken, the particular reason, for the purpose of setting a fair price on the work to be manufactured. He said he understood from the foreman that I objected, and he asked me if it was so, and I told him "Yes," but that it was not an individual or personal matter; and I repeated the action that the molders had taken about meeting, etc. So Maj. Williams then repeated, a little more forcibly perhaps, that he was going to take observations on the floor that morning, and asked me if I was going to submit to it or not. I told him that I could not. He said, "You are discharged." I returned to the foundry on my way to the wash room, and the men, some of them right near me where I went through the door, asked me what was the verdict, and I told them that I was through. The word went along the line and they came to the room to wash and dress, and left the grounds. This is the occurrence, and I don't wish to state anything further on this matter.[27]

Colonel Wheeler's version of the incident differs only in detail. It appears that he did not immediately read the molders' petition — or "ultimatum," as he termed it — when Williams brought it to his office, but gave Williams his instructions under the impression that the difficulty concerned only one man, who had refused to obey orders. In his own words, "The strike of the molders . . . was announced to me before I had scarcely finished reading the ultimatum." [28] On learning of the strike, he appears to have concluded without hesitation that it had been instigated by the union. According to Wheeler's version of the story, Cooney told Major Williams that he personally had no objection to being timed, but that "the organization" objected.[29] With these minor qualifications, there is no dispute as to what happened.

So began the strike at Watertown Arsenal — the spark that was to reignite labor's crusade against the Taylor system and that was finally, in 1915, to result in the outlawing of time study and premium payments in all work done on government contract. Few strikes can have begun with less advance planning or for more seemingly trivial reasons. When the molders came to work that morning, they had had no intention of striking. No one, so far as we can tell from the surviving evidence, had called the strike or ordered the men to leave work. When Cooney was discharged, they simply quit their jobs and walked out. The strike was a fact before anyone realized it.

But were the men really on strike? They at first denied it. At the initial inquiry ordered by General Crozier, the committee of the molders formally objected to the use of the word "strike" in connection with their action, claiming that the molders had left work pending investigation and settlement of their grievances by higher authority.[30] They had not given up their jobs; they were surprised and resentful when, calling at the arsenal a few days later to pick up their personal tools, they were forbidden to enter and made to wait at the gates until their belongings were brought out to them.[31] Such treatment came as a shock: for the first time they realized that their jobs were not necessarily theirs to return to when they wished.

As far as Colonel Wheeler was concerned, the men had given up their jobs. Interpreting his first duty as that of maintaining the foundry in production, he instructed Larkin, the foreman, to find substitute molders as quickly as he could — a move which was only partially successful, as Larkin's sympathies were entirely with the strikers. If the men formerly employed in the foundry wanted to return to work under the same conditions as had existed when they left, Wheeler would consider their request; but he gave no promise that he would re-employ them, nor did he suggest that any compromise was possible on the disputed issues of time study and premium payments. Reporting to General Crozier on the strike, he stated flatly that time studies were indispensable, that they would be insisted on forcibly if necessary, and that, if the arsenal's regular molders objected, "there will be no unusual difficulty in obtaining competent employees willing to work under these methods." [32] As Wheeler saw the matter, men who went on strike gave up all claim to the jobs that had previously been theirs.

The officers of the International Molders' Union were likewise in no doubt that a strike had occurred, and a serious one. The Boston local (No. 106) of the I.M.U. first heard the story on the afternoon of August 11. The immediate reaction was to send out a letter to all Massachusetts Senators and Congressmen, over the signatures of the striking molders, asking their assistance in securing a "fair deal" from the Ordnance Department.[33] On the 14th the secretary of the local wrote to Frank Morrison, secretary of the A.F.L., enclosing a copy of the molders' letter and asking him to use his influence to bring about an early settlement.[34] Both of these letters presented the issue in the same terms: the molders had been required to work under objectionable conditions; they had submitted a protest to the commanding officer, which had been ignored; what they wanted was a fair hearing of their grievances and an investigation by some higher authority.

From International headquarters John P. Frey was sent to Watertown to investigate. He talked with the local union and found it in enthusiastic support of the strike. At the arsenal, where he found the gates patrolled by soldiers with fixed bayonets, he talked with Wheeler, who impressed him as "highly incensed" at the "audacity" of the molders in daring to strike against the government.[35] Frey tried to persuade him to reconsider the matter, arguing that much of the work done in the foundry was not suitable to the application of Taylor methods, whatever might be the case in foundries where the jobs were more repetitive. But he could make no headway. He returned to Washington and submitted his report. Within a few days the Executive Board officially sanctioned the strike.

To the officials of the I.M.U. the issue had embarrassing aspects. Strictly regarded, the strike was in violation of the union's constitution: the molders had gone on strike without even informing the local in Boston; no attempt had been made to negotiate a settlement prior to striking; and no report had been submitted beforehand to the Executive Board.[36] Ordinarily the board would have refused to sanction the strike, and the molders and the local would have been sharply reprimanded. This, however, was a strike against the government — a most unusual event which called for special treatment — and the cause of the strike was a matter on which the Executive Board had already expressed itself in positive terms. Had not the principle been laid down in 1907 that strikes should be authorized whenever a molder was discharged for refusing to work under humiliating conditions or to submit to stop-watch time study?[37] Whether or not the Watertown molders had followed the proper administrative procedure, the issue was not one on which the Executive Board could afford to equivocate. As a strike against the government, the dispute was certain to be closely watched by the union's members, by other elements of organized labor, and by the public. There was no alternative but to give the striking molders official support.[38]

This did not mean, of course, that attempts to mediate between the parties to the dispute were abandoned. Uncompromising though both sides might appear, in reality nobody wanted the strike to continue. The Watertown molders' only demand was that their grievances should be given a fair hearing and that the matter should be investigated by some higher authority. Colonel Wheeler was having little success in recruiting new molders, particularly of the skill and experience he considered desirable; and besides, to have precipitated a strike was no credit on an officer's record. The senior officials of the molders' union disapproved of strikes against the government, and in particular of one that had occurred without their prior knowledge or approval; in terms of the over-all campaign against the Taylor system, the fact that a strike had occurred was enough of a talking point. If it continued, the union cause was unlikely to benefit and might even be injured. To General Crozier, a quick settlement was imperative; conscious of the imminent possibility of intervention by the Secretary of War or by Congress, he wanted the molders back at work immediately. If they thought an investigation desirable, he was not inclined to deny it — provided that the Ordnance Department did the investigating.

The matter was easily arranged. General Crozier instructed his aide, Lieutenant Colonel Thompson, to proceed to Watertown and make a thorough investigation, taking testimony not only from the molders but also from Wheeler, Williams, Barth, and Merrick. Possibly Crozier had already talked with Frey; possibly he had been given an understanding that the men would return to work if their grievances were investigated; the record is not clear on the point. Wheeler at Watertown told the molders that an investigation was under way and that, if they chose to return to work, their jobs were open, an offer that specifically included Cooney. The molders took him at his word and returned to work on August 18 — as if (as Wheeler was to put it later) "they were returning from an ordinary vacation."

They had been on strike just a week. What had they accomplished? Was the situation on August 18 different in any significant respect from the situation seven days earlier? Wheeler thought that it was not, that the men had returned to work under the same conditions as when they left. He intended to go ahead with time studies and premium payments as if nothing had happened — and, as soon as Crozier gave him permission, he did so. Lieutenant Colonel Thompson's inquiry was, he thought, a mere formality; he had done nothing that he regretted, and he confidently expected that Crozier, when he learned the facts, would support him. In this he was not disappointed.

The molders were far from feeling that they had been defeated. They thought that Wheeler had promised that there would be no more time studies until the investigation was completed. Wheeler later denied that he had made any such promise, though he did compromise to the extent of agreeing that "an experienced molder" should be present whenever a time study was taken.[39] Of greater importance, the molders knew that they had won their principal demand: a full investigation of their grievances. Furthermore, their problems were no longer a private matter. Their union was aware of the situation and had backed them up. They knew that their return to work was not the end of the story, but the start of a new chapter — one in which the scene of battle was to shift from the floor of Watertown foundry to the floor of Congress. For the strike at Watertown Arsenal had dramatized labor's hostility to the Taylor system in a way that no previous act of resistance had done. If Wheeler thought, on the morning of August 18, that the victory was won, he was in error. The attempt to install the Taylor system in the government arsenals was in much greater jeopardy after the strike than before it.

One of the first reactions to the strike was the search for a scapegoat. Blame could be variously distributed, depending on one's interpretation of what had happened and on one's desire to evade personal responsibility. Analysis of the various suggested culprits sheds considerable light on the attitudes of the parties involved.

Carl Barth, who had not been present at the arsenal when the strike occurred on August 11, came to Boston on the following day in connection with consulting work he was then doing for the Forbes Lithograph Company. As was his custom, he stayed with Major Williams and his wife. Learning of the strike, he immediately got in touch with Frederick Taylor, sending him a letter in which expressions of altruistic sympathy were neatly combined with assertions of personal innocence. The molders had struck, he wrote, "because of some time study undertaken . . . by Mr. Merrick." The blame lay mostly with Colonel Wheeler and Major Williams, however, because instead of adhering to the original plan of putting some of the molders on premium without time study — to which both Taylor and Barth had agreed — they had got Merrick to go over to the foundry and do some time studies there, and "that turned the devil loose." The whole thing had been unwise and unnecessary: the stop watch should never have been taken into the foundry, since Barth had done no preliminary work there. The tragedy was that Taylor would be blamed. "The enclosed clipping [from a Boston newspaper] will show you that in the eyes of the world, the Taylor system is responsible for the trouble, while the fact is that the real Taylor System man at the Arsenal has never as much as been inside the foundry door. . . . I can't tell you how bad I feel about this matter for your own sake, for in the eyes of the world I fear you can never again say that you have never caused a strike. We will both be found guilty, though neither of us had had a thing to do with it." [40]

Barth's testimony at Lieutenant Colonel Thompson's inquiry followed similar lines: the fundamental cause of the strike was the attempt to make time studies without adequate preparation. "It is my personal conviction," he stated, "that if time study in the foundry had not been attempted without the usual precautions the prevailing sentiment would have been found out during a conversation with the first man approached. In this case I believed the result would have been that no time study would have been undertaken. The result might have been that the authorities at Watertown on discovering this sentiment might have declared a lockout, but I do not see how it would have been possible for a strike to have occurred." [41] Barth's bluntness in placing responsibility on the officers led to some deterioration in his formerly cordial relations with Wheeler. Despite his promise to Taylor to "keep as mum as it is possible for a Barth to do" and to remember that "an officer can't make a mistake," [42] within a few weeks he was congratulating himself on testifying at Lieutenant Colonel Thompson's inquiry "just as the Spirit moved me . . . between my testimony and my report, I think you will realize that there is not much ground on which to expect me to 'share in the responsibility for the strike.' Not much!" [43] And by late September it was necessary for him to assure Taylor that Wheeler seemed to have got over "the unpleasantness of our last meeting" and had not forwarded to General Crozier, as he had threatened to do, copies of Taylor's letters to Barth "so that the 'Chief' could see 'how nasty you and I had been in the matter.' " [44]

Barth's analysis of the causes of the strike overlooked the fact that the molders, in their original protest, had objected not only to time study but also to the earlier experiments with the straight premium plan, of which Barth had approved. The document submitted by the molders to Colonel Wheeler had referred to the "unsatisfactory conditions" that had existed in the foundry "for the past week or more," not to the one day on which time studies had been taken. Neither Barth nor Taylor could evade some responsibility for permitting Wheeler and Williams to install incentive payments in the foundry. Taylor had actually encouraged the step, despite General Crozier's misgivings. Barth had not opposed it, his only proviso being that everyone should understand that what was being done was no part of the Taylor system. What he intended this to mean is not clear; certainly the molders made no fine distinctions of this sort, nor did anyone take the trouble to inform them.

Should Barth have left the foundry insulated for so long from the reorganization of the arsenal? From the beginning the officers had made it clear that, in their view, the foundry was the department most in need of attention. Relevant here are Barth's own lack of *expertise* in foundry management and his obsession with machine-tool operation and the utilization of high-speed steel. It was in the machine shop that the most dramatic immediate results could be achieved. Yet the delay in turning to the foundry was both unsound and imprudent — it was impossible to regularize the flow of production in the machine shop before the output of castings from the foundry had been brought under control, and it left the way open for the kind of amateurish experimentation that Wheeler and Williams undertook. At least Barth should have brought the foundry under the jurisdiction of the planning room, preferably by upgrading the foundry foreman to a position on the planning room staff, and made sure that the molders realized from the start what was intended and why it was necessary. Neither of these steps was taken; the result was the complete breakdown of communication between management and molders that characterized the weeks preceding the strike.

What of Dwight Merrick? As the only other official representative of the Taylor system at the arsenal, surely his explanation of the strike was relevant. Just the contrary, it appears. No one was much interested in Merrick's analysis of what had happened. Nor was Merrick inclined to be pushful, for there were indications at one stage that he might find himself cast in the unwelcome role of chief culprit. His actions had precipitated the strike, had they not? Even General Crozier, doing his best to be open-minded, could not resist asking Taylor whether he thought that Merrick had been "altogether as tactful as he might have been in his way of going about time study in the foundry." [45] Taylor, though feeling personally that if tact were the criterion Williams had a good deal to answer for, was not inclined to spring to Merrick's defense. Barth, however, showed the most obvious inclination to throw his colleague to the wolves. His testimony on the subject at Lieutenant Colonel Thompson's inquiry has an almost contemptuous ring to it. "Mr. Merrick, the expert time study man engaged at this Arsenal on the recommendation of Mr. Taylor and myself, a man, however, whom neither Mr. Taylor nor I look upon as in any way competent to prepare the way for time study or to install the Taylor system." [46] This was consistent with Barth's general attitude to time study, and with the tendency of the Taylor group as a whole. The time-study man's job was, in their view, almost mechanical, demanding only the mastery of some simple skills and requiring the exercise of little judgment or creativity. Time study was important, but the time-study man was not. Carl Barth's later testimony before the Industrial Relations Commission makes the point clearly. "As to the time study," said Barth, "I do not know about it except the theory. I cannot use the stop watch. I wish I could, but life is too short. I turn that over to inferior men; the most important part of the system I turn over to inferior men." [47] Barth was not above deliberate paradox, but his statement nevertheless reflects a fault in the thinking of the Taylor group that later critics were to castigate, and with good reason. [48] Taylor and his disciples — with the notable exception of Henry Gantt[49] — failed to appreciate how strategic was the role of the time-study man. They regarded it as a job that any intelligent clerk could perform, failing to realize that in the eyes of the worker the whole Taylor system came to a focus in the personality and behavior of the man with the stop watch.

Frederick Taylor's attitude to the strike was not very different from that of Carl Barth, though he was inclined to place more responsibility on the unions. He believed that the strike demonstrated the folly of trying to secure quick results. "This ought to be a warning," he wrote to Morris L. Cooke, "not to try to hurry task work too fast." [50] Agreeing with Barth that the strike had been "entirely unnecessary," he expressed the hope that "it will conclusively show all of the government officers the necessity of invariably preparing the ground with a lot of preliminary work before doing anything which directly affects the men." [51] In answer to General Crozier's suggestion that Merrick had shown a certain lack of tact, he stated flatly, "No time study whatever should have been undertaken in the foundry. . . . I have told you time and again that . . . it is only after a year or two of continually harassing men and making them change their ways in minor matters that it is safe to begin on time study and those steps which lead to task work." [52] The strike, as Taylor understood it, suggested no revisions in the doctrine or procedures of the Taylor system; it merely emphasized the dangers of taking short cuts.

In asserting that no time study should have been undertaken in the foundry at that particular time, Taylor was correct. In implying, however, that the officers of the Ordnance Department bore sole responsibility for the departure from the strict prescriptions of Taylorist doctrine, he was conveniently forgetting his own advice of a few months earlier. For had he not, in June, suggested to Crozier the advantages of giving the workmen "a sort of advance benefit" by making "a rough guess" at some of the tasks? By suggesting such a procedure, Taylor had implicitly sanctioned unorthodoxy on the part of others. Taylor's antipathy to short cuts was only skin deep. Less than a year later, in July 1912, he was again urging Crozier to put as many as possible of the Watertown employees on the premium system "by hook or by crook . . . while the Taft administration is still in." [53] This hardly sounded like the discreet, methodical, step-by-step extension of the system that Taylor in his public statements recommended.

But Taylor, even when urging that the men be given some advance benefit in the form of a premium, never recommended the introduction of time study without extensive preparation. And in asserting that if the stop watch had not been brought into the foundry the strike might have been averted, he may have been correct. The molders had resented the premium plan without time study; but against that kind of attempted stimulation they could easily defend themselves without a strike, as Wheeler and Williams had discovered. Taylor was not above blaming the officers for lack of diplomacy; like Barth, he believed that Williams had been altogether too hasty in discharging Cooney. But he thought that no attempt to conduct time studies in the foundry at that juncture, however diplomatic, could have succeeded. The reason was simple: when the men went on strike they were, Taylor believed, following the instructions of their union. "I know of no way in which time study can be tactfully made upon union men who do not understand anything about time study except that it is contrary to union rules, and who in advance have agreed to strike in case time study was made, and I understand this to have been the case at the Watertown Arsenal." [54]

Here, for Taylor, was the true culprit: the union. Perhaps the officers had been arbitrary; perhaps time study had been premature. But if the molders had not already been influenced by union propaganda, the strike would not have taken place. Mistakes had been made, but they were serious only because that old enemy, the union, had exploited them. "It occurs to me," he wrote to Wheeler several years later, "that it is entirely possible that the molders at the Arsenal may have thought that they had an opportunity to 'down' our system by complaining of the way it was introduced. The union men are just smart enough to take whatever you may do and turn it to your disadvantage." [55] Publicly he was more positive and specific; the trouble at Watertown, he alleged, was caused in large part by one document: the O'Connell circular, a vigorous condemnation of the Taylor system, and particularly its use by the Ordnance Department, issued by the president of the International Association of Machinists in April 1911. "The objection on the part of the

men in the Watertown Arsenal . . . was largely brought
about by the utterly unjustifiable and mean misrepresentation of scientific management which was embodied in the
circular which was sent out by Mr. O'Connell, the head of
the machinists' union. . . . Misrepresenting is a mild word.
I would like to use a stronger one, but I do not care to burden
the record with it." [56] Why Taylor thought that a pamphlet
circulated among machinists should lead to a strike among
molders he did not make clear; but that the molders had
struck, if not at the command of the union, at least because
their minds had been poisoned by union propaganda, he was
convinced.

Taylor, though far from welcoming the strike, felt that
it was not without redeeming features. If there was to be a
strike, if the hostility between organized labor and scientific
management was to reach the point of open conflict, then,
thought Taylor, it could hardly have occurred under more
favorable circumstances. He applauded Wheeler's adamant
refusal to make any concessions on the key issue of time
study. When the molders returned to work under the same
conditions — ostensibly — as when they left, he regarded it
as victory. Taylor took this position because, as he saw it, a
fundamental issue was at stake: management's right to
information. "It is impossible," he wrote to Barth, "for you
to stand on any better ground than you do in this case. This
strike hits at the very foundation of scientific management,
and if the owners of the company or the government are not
to be allowed to obtain exact information, then scientific
management becomes impossible." [57] That a strike had occurred was regrettable; but now it had occurred, "there is
only one course to pursue, namely, to fight the thing right
straight through." Management's right to know was not to
be bartered away for the sake of temporary peace.

Taylor was raising issues of profound importance. Granted
that the wage contract conveyed authority to one party in

return for a cash payment to the other, what were the limits of this authority? What were the limits to management's right to seek information about the job and its performance? Taylor's answer was simple: there were no limits. Knowledge was necessary for control; since the goal was complete control of the job in all its aspects, no limits could be placed on management's right to know.

No such complexities characterized the reactions of the officers of the Ordnance Department. Their attitude to the strike had two main elements: in the first place, the issue had been one of discipline; and in the second, the men had been misled by their union. On the matter of discipline neither Wheeler nor Williams suffered from doubts. A man had been given an order which he had refused to obey. Therefore he had been discharged. Was authority to be maintained or was it not? Major Williams expressed this point in his testimony before the House Committee on the Taylor System. "If any workman refuses to carry out any definite instructions that I give him I should immediately recommend to the commanding officer that he be discharged." Asked by the chairman whether this principle applied to a case in which a workman was instructed to proceed with the premium system against his wishes, Williams replied, "My statement is completely general." [58] In taking this position — one that came naturally to a professional soldier — Williams was supported by his superiors. Its virtue lay in the fact that it absolved Williams from personal responsibility.

Colonel Wheeler's attitude coincided in this respect with that of Major Williams, but added to it were hints of personal hurt and disappointment. It was as if he felt that the men had let him down, that they should have known that he would not willingly injure them. If they had resented the changes he had made in the operation of the foundry, why had they not told him? Frustrated paternalism showed in

Wheeler's assertions that he understood his men and was always available to listen to their complaints. The myth of the ever-open door received classic expression: "I personally am always glad to see representatives of the employees, and so far as I know I have never refused to receive any employee who wanted to come and see me about any matter which concerned him." [59] In 1915, discussing the matter with Robert Hoxie, Wheeler recalled that he had first come to the arsenal as a young officer in 1895 and that many of the men who were then working there were still on the payroll. "I generally walk through the shops twice a day and talk with several of the workmen each time I go through. . . . I think they all feel warranted in speaking to me about any matter they think ought to be adjusted . . . they do consult with me about a great many things, even about their private affairs." [60] Yet was there not self-deception here? Wheeler may have believed that there was free communication between him and the men, but from the bottom rung of the organizational ladder the view was different. The men remembered, even if Wheeler did not, that a few years earlier they had requested him to put a suggestion box in the foundry; the request had not been granted, and the event acquired a symbolic importance greater than Wheeler could conceive.[61] Frederick Taylor, after his first visit of inspection to the arsenal, had warned Crozier that the officers had too few conferences with the men, that they "didn't see quite enough of the workmen and didn't circulate among them enough." [62] The surviving evidence leaves little room for doubt that communications at the arsenal, at least between different levels of authority, were highly imperfect. For example, rumors circulated among the men before and after the strike that the older employees would all be fired and replaced by younger, unskilled men; that wage rates would be systematically cut as soon as the speed-up got under way; that Barth, lecturing at Harvard College, had said that the Taylor system would work with "apes and gorillas" as easily as with men;[63] and — most fantastic of all — that an entirely mythical organization known as the "Watertown Improvement Association" intended to "build houses all around the arsenal, so that the men [could] go straight from their work into their beds." [64] Wheeler was no more aware of these rumors than in August 1911 he had been aware of the fast-maturing discontent in the foundry.

As Wheeler remained complacent about his understanding of the men and his ability to perceive how they really felt about an issue, so he saw no reason for second thoughts about the wisdom of the actions that had precipitated the strike. The molders in the foundry were soldiering; he would have been derelict in his duty if he had not used any means at his disposal to break up the practice. Justification for this belief was provided, or so it appeared, by the course of events in the foundry in the weeks immediately after the strike. Among the molders hired to replace the strikers was a man named Harry Edgecomb. He was retained in employment after the strike was settled and proceeded to break all records for daily output. By coincidence, as it appears, he was given the job that, before the strike, had been the responsibility of Hendry: the manufacture of molds for packsaddles. Hendry had turned out an unvarying nine molds a day; Edgecomb averaged twenty-four a day, reducing the total cost per mold from $1.17 to 54 cents and increasing his daily earnings to $5.74, compared with Hendry's $3.28. Wheeler estimated that the daily saving to the government on account of this molder alone was $15.12 and quoted the figures as triumphant proof that he had been right all along. A willing molder could easily match the pace demanded of him and increase his wages in the process.[65]

The case of Harry Edgecomb, however, could be interpreted in other ways, and Wheeler does not seem to have appreciated how peculiar the circumstances were. First,

Edgecomb was a Canadian citizen and therefore ineligible for permanent employment at the arsenal. When hired he had given his birthplace as Denver, Colorado, but he undoubtedly knew that this falsehood would shortly be discovered. Secondly, Edgecomb was given, or arrogated to himself, unusual privileges in the foundry. The other molders alleged that he had cursed an officer to his face without being reprimanded, that he was given priority in the use of the crane, that he was permitted to use for his molds sand that other molders had prepared for their jobs, that he omitted to silica-wash his molds as the other men did, and that in general his molds were inadequately reinforced and finished. As the regular molders saw the matter, Edgecomb was a rate buster, a man willing to exploit the advantages of the moment without thought of the longer-term consequences, because he would not be there when those consequences became real. More than that, they believed that Edgecomb had been treated by the officers as they would never have been treated, solely in order that a record time could be set for the job. They resented his presence bitterly. Edgecomb was not one of them; he was an outsider, a transient, a lone wolf, disruptive of all the values and customs of the work group.[66]

Wheeler's attitude was that the men would discover, once the premium system was in full operation, that the Taylor system was as much for their benefit as it was for the government's and that, if the pernicious effects of union propaganda could be counteracted, no further trouble was to be anticipated. General Crozier shared these views. His verdict was expressed in his testimony before the House Committee on the Taylor System: "At the Watertown Arsenal, when this strike was precipitated, I fancy that the authorities had been lulled to sleep by the smoothness with which the operation of installing the system had proceeded. . . . I think it may be possible that somebody, considering that there was some disadvantage to him in the introduction of this system, thought it worth his while to stir up the employees into a state of dissatisfaction with it and opposition to it. I think it appears abundantly . . . that the Government is not hurt, and that the workmen certainly are not hurt; and I think there arises something of an inquiry as to whether anybody is hurt, and if we can find anybody who is hurt, we may possibly be able to find some reason why these efforts are made." [67] And in case the committee was in doubt as to whom Crozier believed to be "hurt" by the Taylor system, he hastened to make it explicit: "In this matter I can find someone who is hurt, and that is the labor leader. We are endeavoring to do something which will amount to out-bidding the labor leader for the favor of the workman." [68]

Here was the official explanation of the strike — an explanation which, originally stated as a possible interpretation, soon hardened into an article of faith. The men were not — could not be — discontented, since the changes that had been made were obviously in their interests. If hostility arose, it could only be because of the agitation carried on by the unions. The strike called for no rethinking of theory or re-forming of practice; it emphasized merely the unfortunate influence exercised by the unions over the mind of the working man.

That some of the unions, notably the molders' and machinists', had taken a strong stand against the Taylor system and had attempted to prevent its use by the government is beyond question. What influence these organized campaigns had on the behavior of the men at Watertown is a different question, and one not easily answered. Hostility to premium and bonus systems of wage payment was nothing new. The International Molders' Union as early as 1887 had gone on record as being opposed to any system of payment under which a man or group of men who reduced by a specified amount the average time required for a job received a bonus in addition to their daily wages; and in 1907, concerned by the increasing popularity of premium plans among employers, the executive board of the union had decided that strikes should be authorized whenever, under a premium plan, molders were discharged for refusing to do more than "a day's work" or to work under "humiliating conditions" or to submit to stop-watch time study.[69] The International Association of Machinists had officially denounced "work by the piece, premium, merit, task or contract system" in 1903, though the evidence suggests that this formal prohibition was widely ignored by the membership.[70] To the extent that the Taylor system was identified with a particular method of paying wages, union opposition was of long standing; but there is little to suggest that formal opposition had any more influence on the behavior of workmen than had, for example, the long-continued and generally ineffective prohibition of piecework by the I.A.M.

The introduction of Taylorism into government establishments had brought a new note of determination and purpose into the statements made by union leaders. For the first time the Taylor system was identified as the specific object of attack, and there were suggestions of political action against it. It was probably this phase of union opposition that Taylor had in mind as contributing to the strike at Watertown, rather than the earlier opposition to premium and bonus systems. The occasion for the outburst was not Barth's work at Watertown, but certain experiments in stop-watch time study carried out by Major Hobbs at Rock Island Arsenal in September 1908.[71] These experiments had not been authorized by the Ordnance Department and were crude in design and execution. Nevertheless they were regarded by the workers at Rock Island and by their unions as the first step in the introduction of the Taylor system, and their reaction was immediate and violent.

The reaction took two forms: first, a more or less spontaneous burst of protest from the employees at Rock Island, channeled principally through the hierarchy of the Ordnance Department up to the Secretary of War; and second, a more slowly developing attack by the unions, notably the International Association of Machinists, culminating in 1911 in successful pressure for congressional investigation. The reaction that stemmed directly from the Rock Island workers was aimed almost exclusively at inducing the Ordnance Department to order the suspension of time study at that arsenal. Described by Hobbs as "pure, simple, unmitigated and unreasoning opposition," [72] it was nevertheless formidable enough to lead Crozier to order the suspension of time study at Rock Island — though not to abandon his plans for installing the Taylor system at Watertown. The reaction of the unions was more inclusive in its goals and methods and aimed ultimately at achieving, by political pressure, the complete prohibition of the Taylor system in government work.

The situation at Rock Island, after Crozier suspended time studies, subsided until the spring of 1911, when the first moves were made to extend to that arsenal the innovations tried at Watertown. This step provoked new and more violent protests, even though no attempt was made to introduce time study or what Crozier called the "stimulation features" of the Taylor system. In the meantime the organizational attack got under way, exploiting the publicity accorded the Taylor system by the Interstate Commerce Commission hearings of 1910. Early in 1911 the Executive Council of the A.F.L. adopted a resolution condemning incentive-payments schemes and calling upon the affiliated unions to resist the spread of the "speeding system." [73] Samuel Gompers, in the *American Federationist*, set about informing his public as to the nature of the Taylor system — a system designed, as he saw it, to reduce men to the status of machines, to speed them up to the maximum possible rate of production, and to reduce their wages by the competition of common laborers.[74] In April 1911, James O'Connell, president of the I.A.M., authorized the issuance of *Official Circular No. 12*, the document which Taylor believed had done so much to crystallize labor opposition.[75] O'Connell's principal purpose was, like Gompers', to inform; but he also called for action. The adoption of the Taylor system by the Ordnance Department would, he claimed, be a staggering blow to labor; the danger was acute and had to be met immediately by determined resistance. Members were urged to acquire copies of Taylor's *Shop Management*, to study it, and then to write to the Secretary of War, the two United States Senators from their state, and the Congressmen from their district, enumerating labor's objections to the system, protesting its adoption by the government, and requesting the legislators to support any measures submitted to Congress to repress or outlaw the system. The circular made no mention of strikes and recommended only the normal methods of political action.

What the unions wanted from Congress was a prohibition of time study and incentive payments in government work. This was made clear — if any doubt remained — by a statement issued by O'Connell in Washington in November 1911.[76] Asserting that the machinists were unalterably opposed to the Taylor system, he warned that if it were put into operation in government shops one of two things would happen: "Either Congress will enact legislation relieving machinists of the unjust rigors of the so-called 'scientific management' or there will be a cessation of work." The threat was clear; the machinists were prepared to strike against the government if no other means of protecting their interests were found.

What the unions actually got from Congress, at this stage, was much less. It was, in fact, the minimum that the Congressmen concerned could do and retain some hope of reelection. On April 14, 1911, Representative Irvin S. Pepper of Iowa, from the Rock Island district, introduced a resolution (H.R. 90) calling for an investigation of the Taylor system of management. This was referred to the Committee on Labor under the chairmanship of William B. Wilson, who held public hearings on the issue beginning at the end of April. Nothing was said which could not have been predicted by anyone who had followed the controversy of the preceding months. Gompers denied that the Taylor system would increase efficiency in the true sense of the word. Representative Pepper argued that the government should not approve a system which "reduces the laboring man to a mere machine." N. P. Alifas, representing the Rock Island workers, specified the speed-up and the lowering of wages as the consequences of Taylorism that were feared most strongly by the workmen and at the same time regarded as most probable. And O'Connell of the Machinists, alleging that Taylorism had no place for workmen who could think and exercise initiative, said bluntly that "this system is wrong, because we want our heads left on us." [77]

The result was negligible. The committee recommended no legislation and meted out no censure. No further action seemed probable. It looked as if, for the moment, the union attack had stalled and that the workers in the metal trades and their unions would have to learn to live with Taylorism or, at best, oppose it piecemeal as it spread throughout government departments and private industry. In the political field Crozier had his friends. As long as he retained the support of the Secretary of War and could avert congressional action, he had little to fear. Despite O'Connell's threat, a strike against the government was not something to be undertaken lightly.

There was at least one curious aspect to the agitation that the unions had carried on. Watertown Arsenal, and not Rock Island, was the establishment chosen by the Ordnance Department for the initial installation of the Taylor system. If the civilian workers employed by the Ordnance Department really objected violently to Taylorism, was it not to be expected that the strongest protests would emanate from Watertown? Barth had been working there since 1909 and by 1911 had made considerable progress. Yet the surviving evidence does not disclose a single protest from Watertown when the union attack was reaching a climax in the early months of 1911.[78] The petitions to the Secretary of War had come only from Rock Island; the only workers represented at the hearings on Representative Pepper's resolution had been the Rock Island ones. The articles in the union magazines had not even mentioned Watertown. Until the molders' strike in 1911, the labor situation at Watertown, as far as could be discerned from outside the arsenal walls, was as quiet as a churchyard.

If, as Taylor believed, the molders' strike was caused principally by union agitation, it is surprising that, until the strike occurred, the labor situation at Watertown remained as unruffled as it did. Presumably the men read their newspapers; some of them may even have studied the union literature that came their way. They cannot have been unaware of the battle of words going on over their heads. But for all the effect it had on their actions it might as well never have happened. Barth had his problems in the machine shop, but overt hostility on the part of the men was not one of them. Yet the I.A.M. had taken the lead in fighting the Taylor system and had broadcast its opposition with great vigor. As for the molders, John P. Frey, editor of the *International Molders' Journal*, had expressed concern over the Taylor system; but, among all labor writers, he had been notable for the moderation of his remarks and for his refusal to deny that it contained potential for good as well as for evil.[79] If Taylor's theory were correct, Watertown and not Rock Island should have been the focus of rank-and-file resistance, and the machinists, not the molders, should have been the immediate source of trouble.

That an interpretation of the strike that laid primary responsibility on union agitation should appeal to Taylor, Barth, and the officers of the Ordnance Department is perhaps not surprising. That it should become more and more an article of faith as time passed, however, is remarkable, for there was no dearth of evidence to indicate that it was not the pronouncements of the unions but the immediate facts of life in the Watertown workshops that led the workmen at that arsenal to feel and act as they did. The Ordnance Department was at no time between 1911 and 1915 without information as to the sentiments toward the Taylor system of the employees at Watertown Arsenal. These were always positively stated, and they were almost without exception hostile to the Taylor system, in part or in whole. The information necessary for a reconsideration of methods and objectives was available, for those who would listen. Neither Taylor and his immediate colleagues nor the officers of the Ordnance Department would listen. Consequently their understanding of the situation as it existed at Watertown Arsenal (and to some extent at the other arsenals also) increasingly deviated from the facts of the case. The two definitions of the situation — that of the managerial group and that of the employees — grew further and further apart.

The result was, inevitably, conflict — a conflict in which each side found it impossible to believe that the other was in earnest, and in consequence systematically misinterpreted the evidence. In February 1912, for example, hearing of a petition against time study signed by no less than a hundred and eighty machinists and molders at Watertown, Taylor could only comment: "This shows the power of the union leaders over their men. I have not the slightest idea now that these men are really opposed to time study." [80] Later in the same year, with resolutions condemning the Taylor system pending in both houses of Congress, Crozier visited Watertown and reported with pleasure, "I am satisfied that the workmen there now do not want the system discontinued, and I believe that with another year's experience they will be willing to oppose their labor organizations to the extent of saying so openly." [81] And as late as January 1914, with no indication that overt hostility at Watertown had diminished, Taylor believed that "they are kicking because the system is not being introduced fast enough." [82] These beliefs were mistaken. Hostility to the Taylor system among the employees at Watertown did not slacken with the passage of time.

Characteristic of the attitude of Taylor and the Ordnance Department was an exaggeration of the influence of union officials over the behavior and feelings of union members. If it was assumed that union members at Watertown did what their officials told them, and that their attitude toward the Taylor system was a watered-down version of the attitude of the unions, then it became possible to believe that the Watertown employees were — as individuals — content with the system, no matter what the organizations to which they belonged might say. In terms of this logic, the hostility expressed by the unions could be dismissed as not a true reflection of what the employees really felt.

An example of this psychological block was Colonel Wheeler's insistence that the molders had gone on strike on orders from their union. The International Molders' Union had several years earlier given blanket approval to any strikes that might take place as a result of a molder being discharged for refusing to work under humiliating conditions. Conceivably, at the cost of distortion, this might lend some plausibility to an assertion that the strike at Watertown had been ordered by the union. But Wheeler meant specifically that this particular strike would not have taken place if the men had not been given instructions by the officials of their union that they were to go on strike. The only evidence for this belief was Wheeler's own recollection that Cooney, explaining his refusal to submit to time study, had stated that it was not a personal matter but that "the organization" objected. Cooney may have said this, though his version of the incident differs in this respect from Wheeler's; if he did say it, he was correct. But to infer from this that the International Molders' Union had ordered the strike was to leap to a conclusion that the evidence did not justify.

The testimony of the molders on this point is clear, and it is hard to see any reason why they should have lied about it. For example, there is the following exchange between molder Cooney and the chairman of the House Committee on the Taylor System:

Q: At this meeting which you held at which the decision was arrived at not to work if a time study was being made, was it a meeting of your union?
A: No, sir; just the men in the shop.
Q: Were there any others who were not employees of the arsenal present at that meeting?
A: The first meeting?
Q: Yes; the meeting at which this decision was arrived at.
A: No, sir.
Q: Do you know whether any communication, oral or written, had been conveyed to any of the men by any officer of any of the unions, suggesting that this action be taken?
A: No, sir.
Q: Then the action was purely spontaneous on the part of the workmen themselves?
A: Yes, sir.[88]

Wheeler's misinterpretation may have been given apparent support by the report of the incident in the *International Molders' Journal*, which referred to it as a "sanctioned strike." But it had been sanctioned after, not before, it took place. To John P. Frey, reporting in 1953 on his memories of the affair, the most distressing aspect of the strike was that it had occurred "in violation of the molders' constitution" and that neither the local nor the Executive Committee knew anything about it until after the men had walked out.[84]

It was possible to concede that the union had played no direct role in precipitating the strike and to believe nevertheless that the hostility of the men to the Taylor system, both before and after the strike, was the result of union propaganda. This was Taylor's point of view, and in a certain qualified sense it was not without validity. As an explanation of the course of events *after* the strike it was highly plausible. Leadership in the prolonged campaign for a congressional prohibition of time study in government work lay with the unions, particularly the International Association of Machinists and the International Molders' Union. The employees at the arsenals played a supporting, though indispensable, role. They provided the raw material for the unions' campaign, and it was their positive and frequently expressed hostility that the unions exploited. To say this is not to say that, had the unions sat back and done nothing, the hostility of the arsenal employees would have subsided into disgruntled acquiescence. Possibly this might have happened; more probably their resentment would have expressed itself in a different form. In general, an examination of the employees' attitudes, as expressed in personal testimony, leaves one with the strong impression that their grievances were too immediate and particular to be quickly forgotten. This at least can be said: the unions provided the leadership and channeled resentment into a campaign directed toward specific legislative goals; but the objections expressed by the employees to the Taylor system were not primarily a reflection of the organizational pressures of the unions.

In asserting that the strike was the result of union propaganda, Taylor was on less firm ground. And in claiming that the O'Connell circular in particular had precipitated the strike he came close to talking nonsense. The O'Connell circular was addressed to machinists, not molders; it was neither reprinted nor referred to in the *International Molders' Journal;* and there is no evidence that any of the molders at Watertown Arsenal had ever heard of it, far less read it. John P. Frey had condemned any system of management that drove men beyond their normal capacity, and he had stated his fears that, if the Taylor system spread, it would entail the destruction of craft unionism. But he had published nothing that could be described as incitement to direct action. The question may be raised whether anything that had been written in union periodicals about the Taylor system had any effect whatever upon the behavior of the Watertown employees. Not all of them read the official periodicals of their unions, and not all who read understood the relevance of what was written to their own situation.

There is one instance in which the written word does seem to have affected the behavior of the men at Watertown. This was Taylor's *Shop Management.* None of the Watertown employees, when testifying before Lieutenant Colonel Thompson's *ad hoc* committee of inquiry or the later House Committee on the Taylor System, referred to any union publication; several of them, however, referred to Taylor's monograph in terms suggesting that they had read it and had given thought to its contents. Alexander Crawford, foreman of the erecting shop, said that several copies of Taylor's paper had been given to the arsenal "at the time of the installation of this Taylor system" and that his knowledge of the system was based on his reading of one of these copies.[85] And Cooney's report of the molders' discussions on the night before the strike states that they concluded they could not submit to Taylor methods if they were carried out in their entirety "as they are recorded in his book called Taylor's 'Scientific System of Shop Management.'"[86] How carefully

Taylor's article had been read, and how well it had been understood, it is impossible to say; the molders did not understand what Merrick was doing when he made his first time study in the foundry. But there were copies of Taylor's writings available to the men in the arsenal; some of them had been read; and the men's opinions were based in part, at least, on this reading. There is no corresponding evidence as to the influence of the union journals.

One problem should have troubled those who blamed the strike on union propaganda. The most violent denunciations of the Taylor system had emanated not from the molders' union but from the machinists'. If union agitation had stirred the men up against the Taylor system, should it not have been most effective in the machine shop? Why was it that, up to August 1911, there had been no overt signs of dissatisfaction among the Watertown machinists?

Taylor had an answer to this, and within limits it was a sound one. In the machine shop the correct procedure for installing the Taylor system had been followed. Barth had spent two years reorganizing working conditions before time study had been introduced. The routing of work, the respeeding of the machines, the reorganization of stores and maintenance procedures — these and many other innovations had created a climate of change in the machine shop such that, when the first task was set for a man by time study, it seemed only a minor extension of the work that had already been done. In the foundry the sequence of events had been just the reverse. Wheeler and Williams had plunged in, determined first to set tasks without time study, then equally determined to use time study, without any serious attempt to standardize conditions or to get the men into a suitable frame of mind. Taylor had always insisted that a long period of preparatory work was indispensable if an incentive-payments scheme based on scientific time study was to work. The contrasting results of Barth's sophisticated approach in the machine shop and the naïve impetuosity of Wheeler and Williams in the foundry seemed to confirm his thesis.

There were, however, certain elements in the situation that Taylor did not consider. In particular, he seems to have mistaken an absence of overt hostility for contentment. The testimony of the machine-shop employees — at the various hearings after the strike and in their petitions to Congress and the Secretary of War — shows that they were no less stubbornly opposed to the Taylor system than were the molders. The difference between the situation in the machine shop and that in the foundry was not in attitudes; it was in the fact that in one department hostility had led to action, whereas in the other it had not. And for this there were several reasons. If a stand was to be taken against time study, it had to be taken as soon as the stop watch made its appearance in the shop. This demanded a considerable solidarity among the men and the ability to arrive at a consensus, without prolonged dispute, that this innovation had to be opposed immediately. The molders in the foundry possessed this solidarity; as a group they demonstrated an ability to produce leadership when it was needed and to reach a mutual understanding with unusual facility. Cooney did not seize the role of leader; he had it thrust upon him. The informal organization of the foundry was strong enough to produce an effective and vigorous response without outside intervention. Not so in the machine shop.

Why was the reaction in the machine shop so different from that in the foundry? The machine shop was a much larger department; it contained at this time a hundred and forty-six machinists, and numerous helpers and workers of other trades, compared with the foundry's eighteen molders. Size alone made a difference to the effectiveness of informal organization. Then again, the challenge to which the molders responded was an immediate and obvious one, whereas in the machine shop the process of change was more gradual. Furthermore, in the machine shop the formal structure of authority had been changed by the advent of the Taylor system. The master mechanic and the foremen had been removed from their duties on the floor of the shop and transferred to the planning room. To some extent this increase in authority may have reconciled them to the new system of management, although it did not do so completely. At least it insulated them from face-to-face interaction with the machinists and deprived the social system of the machine shop of some of its most important opinion leaders — men who in normal circumstances could be counted upon to indicate, by a word, a gesture, perhaps merely a facial expression, what was the appropriate reaction to any change in the work situation. In the foundry, the foreman, Larkin, received no corresponding elevation in status. Merrick's appearance in the shop may have seemed to him a threat to his authority over the men. Significantly he was discharged shortly after the strike, on the ground (according to Crozier) that he was "not in sympathy with . . . the new system in the foundry at all. He was opposed to it; he didn't want it." [87] Any possibility that the foremen might encourage opposition to the system had been canceled out in the machine shop by a change in formal organization; in the foundry the structure of the situation was in this respect different.

The molders asserted that, whatever might be true of the machine shop, the foundry was not adapted to the Taylor system. The jobs were too varied, and the work was largely done by hand, not by machines. "The men don't generally object to the speeding up of the machine," said one of their spokesmen, "but the speeding up of the molders, a hand operation, we decidedly object to." [88] Any increase in the rate of production in the foundry, they felt, could come only by increasing the rate at which they, not machines, worked. This view involved an underestimation of the increases in productivity that could be obtained by greater division of labor, by a smoother flow of material and orders, and by job simplification, but nevertheless it carried weight. In terms of mechanical equipment the foundry was poorly furnished, as Colonel Wheeler, under cross-examination, was forced to admit. There were two cranes and a few pneumatic rammers, and that was all.[89] Nothing had been done to modernize or re-equip the shop before introducing time study and incentive payments. This suggests that of the two possible avenues to higher productivity — increased investment in mechanical equipment or more systematic job supervision — the officers chose the latter, probably because of its lower initial cash costs. But it also suggests — a point of more relevance here — that the molders, unlike the machinists, had received no demonstration of the immediate and direct effects on productivity that the Taylor system, in its more mechanical aspects, could produce. In a machine shop the impact of Taylor methods on machine-tool operation could be highly dramatic, even exciting to a man who took pride in his craft. The demonstration effect of the correct use of high-speed steel was something of which Taylor and Barth were well aware and not above exploiting if the situation demanded it. Nothing of this kind was possible in the Watertown foundry; here the impact was predominantly on hand processes, and the molders remained unconvinced that the new system could increase production without making them work faster and harder. With no machine time to be cut, the point of attack could only be handling time.

One difference between the machine shop and the foundry — Taylor would certainly have made much of it if he had known — was that the former was far from completely unionized. There was, in addition, a serious problem of dual unionism. Not many years before, the International Association of Machinists had been able to claim 100 per cent membership among the employees in the machine shop at

Watertown Arsenal.[90] Since that time, largely as a result of the inroads made by a rival organization known as the National League of Government Employees, membership in the I.A.M. had slumped badly. In March 1911, District 44 of the I.A.M., covering the machinists employed at Watertown, reported tax receipts of only $1.20. At that time the per capita tax was 5 cents per member, which would indicate that there were not more than twenty-four dues-paying members employed at Watertown Arsenal out of a total of a hundred and forty-six machinists.[91] One result of the Taylor system was to rebuild the membership strength of the I.A.M. at Watertown. In June 1911 a representative of the I.A.M. visited Watertown and addressed the machinists during the noon-hour break. He reported that there was a good deal of unrest and that the news that the men would be put on premium work had "at last aroused them to the necessity of making intelligent organized resistance." [92] Most of the machinists, he added, had formerly been members of the I.A.M. but had affiliated with "some ten-cent organization" (presumably the League of Government Employees). By the end of August, after the molders' strike, sixty-two machinists had reaffiliated with the I.A.M., and the rising trend continued thereafter.[93]

It seems evident that the installation of the Taylor system at Watertown, far from weakening union organization (as Taylor doctrine would have predicted), actually strengthened it. The International Association of Machinists, by its vigorous opposition to the system, was enabled to capitalize on the feelings of dissatisfaction already existing among the Watertown machinists and to rebuild its position against the competition of a rival organization. The initial weakness of the I.A.M. at Watertown supports the assertion that dissatisfaction among the employees was not the creation of the unions but a factor that the unions exploited. It also lends strength to the hypothesis that informal organization in the foundry, where the molders were all members of the same union, was considerably more effective than in the larger machine shop, where the ability of the employees to agree on a common course of action was limited by conflicting allegiances to rival unions. Whether the I.A.M. deliberately entered upon its campaign of opposition to Taylorism with the intention of rebuilding its membership in the government arsenals and Navy yards is an open question. That it had the strength, before the summer of 1911, to incite opposition where none already existed seems highly improbable.

Chapter 5

CONSEQUENCES

The results of the installation of the Taylor system at Watertown Arsenal can be appraised on two levels of analysis at least. We can, to begin with, inquire to what extent the stated objectives of the innovation were achieved. This directs attention to evidence bearing on changes in productivity and costs of production. But we can also examine the unexpected consequences, the results that were not regarded as particularly desirable by the innovators but which conditioned and limited what they were trying to achieve.

The Taylor system was adopted in the arsenal in the expectation that it would bring about reductions in costs of production and in general raise productivity. To what extent were these objectives realized? Something depends on the date at which comparisons are made. When was the installation of the Taylor system complete? In a sense, it never was complete: The Taylor ideal was not fully realized at Watertown — or in any other plant, for that matter — and there is ample evidence that, even when the system at Watertown was regarded by those in charge as fully installed, there were many deviations from the formally prescribed methods. Any date line for comparisons is to some extent arbitrary. Barth paid his last visit as consultant in June 1912, and Merrick left the arsenal twelve months later. By the middle of 1913, therefore, the system was on a self-sustaining basis. This did not mean that it covered the entire arsenal. Work in the foundry was still not controlled from the planning room as late as 1915. If the amount of work done under the premium system is taken as the index, over the whole of 1914 the molders spent 73 per cent of their working time on premium jobs, the machinists 66 per cent, the teamsters 73 per cent, and the toolmakers 23 per cent.[1] Wheeler, in June 1915, referred to the installation of the system in the machine shop as "practically completed"; in the foundry he estimated that about 70 per cent of the work necessary for the completion of the system had been done, in the smith shop about 50 per cent, and in other departments about 35 per cent.[2]

During the years from 1908–1909 to 1914–1915, the evidence suggests a substantial degree of success in accelerating the pace of work. One indication may be found in the premiums earned. Average base rates of pay at the arsenal increased between December 1908 and February 1915 from $2.93 to $3.07 for machinists and from $3.11 to $3.46 for molders. If premiums earned are included in the calculation, however, the average daily earnings of a machinist between November 24 and December 23, 1914, were $3.76, while the molders during the same period earned an average of $4.56 a day. The average premiums earned by machinists at this time were 22.18 per cent of the daily rate; for molders the corresponding figure was 31.62 per cent; and for toolmakers, 18.95 per cent.[3] Table 1 shows average daily earnings on day

TABLE 1

Average Daily Earnings at Watertown Arsenal, 1914

	Day work	Premium	Hours on premium as per cent of entire working time
Blacksmith	$3.18	$3.87	64.62
Carpenter	3.01	3.96	19.36
Machinist's helper	2.08	2.64	14.72
Molder's helper	2.10	3.78	7.30
Machinist	3.07	3.80	66.00
Molder	3.44	4.42	72.93
Teamster	2.05	2.61	92.84
Toolmaker	3.36	4.22	22.49
Toolsmith	3.28	3.85	7.01

work and on premium for selected trades during the year ending December 31, 1914. Wheeler reported that practically all the workers put on premium jobs began to earn premiums immediately and that failure to earn premiums was practically unheard of except in the case of machinists, where a failure rate of approximately 4 per cent seemed to be normal.[4]

Statistics showing changes in labor productivity after the installation of the Taylor system are for the most part not available. Certain examples of reductions in the cost of production of particular products will be presented shortly. For the foundry, Table 2 shows, in the form of monthly

TABLE 2

Output of Castings per Man per Day, 1909 to 1912 Inclusive[5]
(monthly averages, in pounds of metal cast)

	1909	1910	1911	1912
January	502	682	398	629
February	526	659	341	780
March	497	632	574	1004
April	591	590	478	1166
May	556	559	416	1007
June	545	596	328	1137
July	448	363	285	579
August	509	502	299	1038
September	463	520	405	1332
October	740	501	591	1479
November	649	364	632	1320
December	649	445	618	1500

averages, the output of castings per man per day in pounds for the years 1909 to 1912 inclusive. The steeply rising trend on the chart (p. 189) during 1912 contrasts sharply with the practically horizontal trend during the previous three years. (In each year a pronounced drop in output in July reflects the summer vacation.)

These increases in productivity per man were achieved with no significant expansion of the total work force. Sixty-

Watertown Arsenal

Foundry: Output of Castings per Man per Day in Pounds for the years 1909, 1910, 1911, and 1912 (monthly averages)

nine men were employed in the foundry in 1910 and 1911; by March 1915 the total had risen only to seventy-one. There appears to have been a redistribution of manpower within the department: in 1911 there were eighteen molders out of the total work force of sixty-nine, while in 1915 there were only eleven out of a total of seventy-one.[6] The change is to be explained by an expansion of the number of molders' helpers and other ancillary labor. At the same time considerable capital investment took place. Funds authorized for the purchase of new machinery and the modification of old machinery in the foundry totaled $55,370.00 in the fiscal year ending June 30, 1911, compared with only $350.00 in the previous twelve-month period. In the machine shop a similar process was taking place; only $33.25 was spent on the equipment of that department in the fiscal year ending June 30, 1910, compared with $25,795.58 in the following twelve months.[7] Part of the increases in labor productivity is undoubtedly due to larger inputs of capital, rather than to the specifically Taylorist innovations; to separate the effects of the two processes is impossible.

Evidence on the reductions that were alleged to have occurred in costs of production is to be treated with skepticism for a number of reasons. First, the data are drawn from testimony given by officers of the Ordnance Department who were concerned to present the results of their managerial reforms in as favorable a light as possible. The cost figures they reported were probably not deliberately distorted or faked, but there is no assurance that they were representative of a general reduction in costs of production throughout the arsenal. In some cases the officers stated that they were presenting the best examples they could find, or in other words that they were taking a selective sample from all available cases. Secondly, it may be asked whether the examples given proved what the officers thought they proved: alleged reductions in costs of materials were to some extent the result of changes in bookkeeping only. And thirdly, important changes in cost-accounting methods were being introduced at the arsenal through the period over which cost comparisons were made. These changes related specifically to the method of allocating overhead costs. If two cost estimates for the same product are presented, relating to different dates and calculated by a different method of distributing overhead costs, they are not, strictly speaking, comparable.[8] The second figure may be lower than the first merely because the product, at that date, is made to carry a smaller share of overheads. The data that might make possible the calculation of costs of production on a standard basis before and after the installation of the Taylor system are not available.

Not only was the method of allocating overhead costs changed; there was also a redefinition of what costs were to be considered as overheads. When Barth first reported to Crozier on the reforms he proposed to institute at Watertown, he warned him that the Taylor system usually involved a somewhat startling increase in the ratio of unproductive to productive labor and went on to argue that this ratio was in no sense a criterion of efficiency (see above, p. 88). As the terms were used at that time, productive labor referred to labor that could be directly and unambiguously assigned to specific jobs; unproductive labor — a most inappropriate phrase — referred to labor regarded as part of the overhead costs or burden of the establishment. A machinist's helper or a foreman might be classified as unproductive labor, merely because it was difficult to keep track of the time he spent on each particular job.

As a result of the installation of the Taylor system there occurred a marked transfer of labor from the productive to the unproductive category. This entailed, purely as a matter of bookkeeping, a decrease in the direct labor costs of each job and a corresponding increase in overhead costs. To the extent that this took place, reductions in labor costs could be claimed that had no relationship to improvements in labor productivity. The point may be illustrated by Table 3, showing how the workers in the arsenal foundry were classed, for cost-accounting purposes, as productive and unproductive in the years 1909, 1910, and 1911. Between 1909 and 1910

TABLE 3

Classification of Labor in Watertown Arsenal Foundry
(June 30, 1909, 1910, and 1911) [9]

Occupation	Numbers employed		
	1909	1910	1911
Productive:			
Chippers	12	8	11
Coremakers	2	2	2
Molders	27	22	18
Molders' helpers	31	11	11
Molders' apprentices	1	2	2
Totals	73	45	44
Unproductive:			
Foreman	1	1	1
Storekeeper	1	1	1
Assistant storekeeper	1	1	1
Annealers	2	2	2
Cranemen	2	2	2
Molders' helpers	0	12	12
Melters	4	4	4
Skilled office laborer	0	0	1
Chipper (leading man)	1	1	1
Totals	12	24	25

a sizable reclassification took place. This in itself would produce reductions in direct labor costs. It is not suggested that these changes were made purely to generate favorable reductions in direct labor costs; but the effect of such reclassifications must be borne in mind when analyzing the impact of the Taylor system on costs of production.

With these considerations in mind, some of the examples of reductions in cost that were presented as demonstrating the effect of the Taylor system of management at Watertown Arsenal may be examined. Not all examples are given in equal detail. It will be convenient to begin by discussing a single case at length, as illustrating both the general tendency of the evidence and some of the problems of interpretation that arise.

Case 1 was presented by Colonel Wheeler in his testimony before the House Committee on the Taylor System. It relates to sets of parts for the modification of twelve-inch mortar carriages. Ninety-six sets of these parts were ordered and manufactured before the installation of the Taylor system; five orders of forty sets each (two hundred in all) were manufactured after the installation had begun. Wheeler presented the following figures as showing average costs of production per set before and after the introduction of the new system of management.

Case 1: Sets of Parts for Modification of Twelve-inch Mortar Carriages[10]

	Under old system (1)	Under Taylor system (2)	Absolute reduction	Reduction as percentage of (1)
Labor costs	$ 483.85	$275.00	$208.85	43.13
Shop expense costs	335.88	332.00	3.88	1.16
Material costs	785.02	362.00	423.02	53.89
Total	$1604.75	$969.00	$635.75	39.62

It can hardly be denied that, prima facie, this is an impressive showing. Material costs per set were reduced by over one half, labor costs by over two fifths, and total costs by almost two fifths. Let us analyze the various categories more closely.

Labor Costs: This figure comprises the total cash payments made to all workers directly engaged in operations on the job in question. A reduction in labor costs of two fifths means that the total time spent by such workers on this job was reduced by more than two fifths, since the workers were paid premiums based on the time saved and these premiums were included in labor costs. The reclassification of workers from the productive to the unproductive category would have a definite effect on the figures given for labor costs. Other obvious factors tending toward a reduction in labor costs are the incentive-payments system, giving the worker a direct financial incentive in completing his task in less time than he was accustomed to take for it under the day wage system, and the respeeding of machine tools and other strictly mechanical reforms instituted by Barth and Merrick. Less obvious, but perhaps no less important, was the introduction of more precise methods of estimating the time actually taken on jobs and of systematic procedures for expediting the transition from one job to the next. Both these factors — the one a result of time study and the use of job cards, the other of the introduction of a planning and routing system — would tend to reduce direct labor costs.

Major Williams emphasized the importance of more accurate timekeeping in his testimony before the House Committee. "The timekeeping," he stated, "which determines the direct labor, is very much better than it used to be. We used to have time clocks at various places in the shops, and the foremen made out the job cards themselves and handed them to the men. The men rang in and out on these job cards which were prepared by the foremen and you had a double chance of inaccuracy, and the foremen's job cards were not always accurately prepared; sometimes material would be inadequately described and, perhaps, even a wrong charge number used. At other times men undoubtedly kept at work on a job card long after the job called for by that card was completed. The method now determines, in my opinion, very much more accurately the amount of labor that is put on a given order than the former method did, so that the increased accuracy comes both in labor and materials." [11] In other words, more exact timekeeping not only resulted in more precise measurement of direct labor costs; it also, by making possible stricter managerial control over the workmen, tended to reduce those costs.

With reference to the routing system, Colonel Wheeler commented in 1915 that: "There is less waiting around. Each workman now gets the work that he is better qualified to do than before; there was very much of a scramble before; the first man that finished a job, in a great many cases, got work that he was not well fitted for. He got it because the foreman was so overburdened with clerical and other work that he was glad to assign a job to anybody to get it going." [12] Similar economies in direct labor time resulted from the new system of issuing stores and materials. Again contrasting Taylor methods with the old system, Wheeler stated that: "It was frequently the case . . . for a machinist to get his order to do a certain job, and then necessary for him to get a stores truck and a helper and wander around the shop to seek and obtain the materials he needed to do the job. We now, under this present arrangement, have a different instrumentality by means of which the stores are delivered to him and he has no responsibility whatever. . . . No order can be given to a man to do a job until the material has been delivered. That saving cannot be evaluated. The delay in getting material and the time consumed in looking for it was, under the old way, considered as part of the job." [13]

When we consider the total impact of all these factors, the large reduction claimed in direct labor costs becomes more understandable. It is difficult to estimate the relative importance of each factor. If one may judge by the weight attached to each in the testimony of the officers in charge, the more systematic method of distributing tools and material to the worker before the job was begun seems to have been of major importance and may even have produced greater savings in time than the incentive wage system on which so much controversy centered. It was not only that functions such as hunting for tools and materials which, as Wheeler put it, had previously been considered "part of the job" and left to the initiative of the worker were now converted into a routine feature of shop organization; from the cost point of view, the important change was that the time taken in performing these functions was no longer assessed against the job as direct labor costs. The functions were now assigned to individuals who were responsible for nothing else. Such individuals, being unskilled, were cheaper, and the cost of employing them became part of the shop expenses or overheads.

Shop Expense Costs: In the example under consideration, the reduction claimed in shop expense costs was only 1.16 per cent. Two opposing tendencies were at work here: an increase in the aggregate shop expense burden, and a decrease in the length of time for which each job carried its share of this burden. In this case the two tendencies seem to have nearly canceled out.

The aggregate overhead costs of the arsenal increased as a result of the installation of the Taylor system. The additional expenses incurred for the regular operation of the Taylor system at Watertown Arsenal were estimated in 1912 at approximately $24,000 per year.[14] Total overhead costs in the fiscal year ending June 30, 1909 — including shop expenses in the strict sense and also the "Arsenal burden factor," comprising such things as cost of leaves, inspection, power, and planning — were $236,737.06. One may tentatively conclude that the installation of the Taylor system involved an increase in overhead costs of a little over 10 per cent. This was reflected in increases in the shop expense ratios of the foundry and the machine shop from 160.65 per cent and 90.65 per cent respectively in 1909 to 241.15 and 116.75 per cent in 1911.

These increases were due partly to the reclassification of workers, partly to the fact that management was now performing functions which prior to the installation of the Taylor system it had not performed, and partly to a rise in expenditures on such auxiliary services as heat, light, and power. Any decrease in direct labor costs due to reclassification of workers would be offset by a corresponding increase in the shop expense charge. The change would be purely a matter of bookkeeping. A decrease in direct labor costs due to time economies in production, on the other hand, would not necessarily be offset by a rise in the shop expense charge, even if the shop expense ratio increased considerably. A cut of, say, 10 per cent in the time taken to machine a casting meant not only a reduction in direct labor costs; it meant also a cut in the time during which the job had to bear the shop expense ratio. Whether the shop expense charge for that job increased or decreased would depend on whether the rise in the shop expense ratio was more or less than sufficient to offset the fall in the time taken to machine the casting. An increase in the volume of production thus meant that any given aggregate of shop expenses could be distributed over a larger number of units of product, reducing the load of overhead costs on each unit.

This is an elementary point, but it must be understood to grasp the circumstances under which the Taylor system would or would not result in a reduction in total costs. Everything depended upon whether or not the time economies achieved in production outweighed the rise in overheads. An example will clarify the point. Colonel Wheeler estimated that, on the average, a job could be done 2.64 times as fast under the Taylor system as it had formerly been under the day wage system. Let us assume that this figure is correct and that a machinist whose wage rate was $3.28 a day could turn out a certain job under the Taylor system in one day. Under the old system it would have taken him 2.64 days. The shop expense ratio in the machine shop before the installation of the Taylor system was 90.65 per cent; under the Taylor system it was 116.75 per cent. A simple calculation shows that, under these conditions, time economies vastly outweighed the rise in overheads.

Under the old system:

Direct labor costs ($3.28 × 2.64)	$8.66
Shop expenses ($8.66 × 0.9065)	7.85
	$16.51

Under the Taylor system:

Direct labor costs ($3.28 × 1.00)	$3.28
Premium (one third of day wage)	1.09
Shop expenses ($3.28 × 1.1675)	3.83
	$8.20

If Wheeler's estimate of the time economies achieved under the Taylor system is correct, savings in total cost would result at any shop expense ratio short of 370 per cent.

In Case 1, parts for the modification of twelve-inch mortar carriages, it is possible but not certain that some of the reduction in shop expense costs claimed for the later orders may have been due to the new system of distributing such costs. By 1911 the machine shop was no longer applying a standard shop expense ratio to every job that went through; the shop expense costs for the later batches of this order were calculated, in the machine shop, by machine-hour and man-hour rates (see above, p. 118). The flat shop expense ratio quoted above for the machine shop in the year 1911 had to be specially calculated for presentation to the House Committee on the Taylor System; it was no longer in use at the arsenal. What part, if any, of the reduction in shop expense costs can be ascribed to this factor is not calculable; there is no method, in the absence of the original data, by which cost figures calculated on the new basis can be recomputed to show what costs would have been if calculated on the old basis. This might be possible if only one department were involved, for in that case the flat shop expense ratio for 1911 could be applied to the direct labor cost figure. But this order went through all the shops, and there is no means of discovering what proportion of the direct labor costs should be allotted to each.

Material Costs: In Case 1, the installation of the Taylor system is alleged to have resulted in a saving in cost of materials of 53.89 per cent, or just under two thirds of the total reduction in cost. This is at first sight surprising. There is no reason to expect that the Taylor system would result in any economies in utilization of raw materials.

The explanation lies in the elimination of the inefficiencies that had characterized the older methods of ordering and requisitioning materials. Before the installation of the Taylor system these were the responsibility of the foremen. The testimony of the officers in charge suggests that it was normal for the foremen to overestimate their raw material requirements, to submit duplicate orders if material was slow in arriving, and to order in relatively small quantities. There was no systematic check upon the judgment of the foremen or upon duplicate orders. Under this system, what was charged against a job as material costs was not the material actually used on that job, but that ordered. The chronic tendency to inflate and sometimes to duplicate requisitions meant that the material cost of each job was typically greater than the cost of the materials used. The arsenal was not authorized to hold inventories of materials, so that anything purchased under a certain fabrication order number was charged against that order, whether it had been used or left lying around the shop.

The economies in material costs resulted from remedying the obvious defects of this system, in particular from the innovation of charging to each order only the material actually used for that order. These were not economies in utilizing materials; they were economies in requisitioning them and in allocating their costs.

Two principal changes were made, the first of which involved the securing of authority for the arsenals to hold inventories of materials. In 1915, Colonel Wheeler referred to this change as having been made "very recently." It enabled material to be bought in larger quantities, with benefits in prices paid and in the speed with which material could be issued to the shops. The second change was the formation of the engineering division, which took over from the foremen the function of preparing bills of material and thus diminished the risk of overestimating requirements and duplicating requisitions.

The question may be raised whether these two innovations are properly regarded as part of the Taylor system. Taylor and Barth may have approved of the new inventory policy, but there is no evidence that they suggested it. The engineering division was instituted by Wheeler on his own initiative before either Taylor or Barth had visited the arsenal. There can be little doubt, however, that these changes or something similar would have been made as part of the Taylor system if they had not been made previously and that their beneficial effects were accentuated by other innovations for which Taylor and Barth were responsible — by the formation of a central storeroom with efficient procedures for ordering and issuing materials and by the extensive use of job cards, which laid down precisely what materials were to be used for each step in the process of manufacture. Nevertheless, these reforms instituted at the arsenal prior to the advent of Carl Barth were chiefly responsible for the reductions in material costs, and of them the removal from the foremen of the function of estimating and requisitioning material requirements was probably the most effective.

It is possible to argue that these savings in material costs were to a large extent mere matters of bookkeeping. They were reductions only in a certain sense — the sense in which you have made a "reduction" in the height of a man when, having guessed that he is six feet two inches tall, you take a tape measure and find that he is actually only five feet eight. From this point of view, a true reduction in material costs would imply that the arsenal got more output per dollar's worth of raw material input. There is no evidence that this happened. The later estimates of material costs were lower than the earlier simply because the earlier were bad estimates.

From a different point of view, however, there is nothing

illusory about these reductions. If the commanding officer of the arsenal had been asked in 1907 to quote a figure for the material costs of manufacturing one set of parts for the modification of mortar carriages, he would have set the figure at $785.02. And if a decision had had to be made at that time as to whether the arsenal should manufacture these parts or contract out for their production to a private concern which had quoted a lower price, the decision would have been made on the basis of the estimate. In 1911, if a similar decision had to be made, the arsenal's figure would have been $362.00 and the decision would again have been based on that figure. In the first case the arsenal might not have secured the order; in the second it probably would have. In this sense the question of the accuracy of the method of estimating material costs is irrelevant. The true cost figure is the figure accepted as true by the individuals concerned, for purposes of action. From this point of view, there is no question but that substantial reductions in material costs were achieved.

It would be possible to argue at length about the reality of these savings in material costs, but the question is a semantic one. They were achieved by the adoption of new techniques of cost accounting, not by any change in production-line jobs. This is not to deny that the tightening up of accounting methods may have had important side effects, such as the paying of lower prices through quantity purchasing, or a reduction in the average time that elapsed between authorizing an order and getting it into production, or greater accuracy in estimating future costs — a factor of some importance when it came to drawing up production schedules in terms of annual appropriations. These benefits, however, are not to be measured by the figures presented as showing reductions in material costs. As Colonel Wheeler correctly and succinctly stated, "It is not a saving really in material; it is a saving in material as far as these orders are concerned. These orders paid for that amount of material." [15]

The available evidence suggests that, of the total reduction in costs claimed for the Taylor system, almost two thirds resulted from decreases in material costs. That Case 1 is not atypical is demonstrated by the following cases put forward by Colonel Wheeler in 1915.[16] These relate to cost of material per unit of product only.

Case 2: Ten-inch Disappearing Gun Carriage, Model of 1901

Expenditure Order No. 2010 (August 1, 1905)	$11,758.92*
Expenditure Order No. 4538 (August 13, 1908)	6,275.40

Case 3: Fifteen-pounder Barbette Carriage, Model of 1903

Expenditure Order No. 2434 (February 1906)	$ 334.63
Expenditure Order No. 3185	337.76
Expenditure Order No. 3904	344.83
Expenditure Order No. 5267	249.76
Expenditure Order No. 6637	207.74
Expenditure Order No. 8423 (June 1912)	216.53

Case 4: Six-inch Disappearing Gun Carriage

Expenditure Order No. 3960 (August 1907)	$ 4,114.16*
Expenditure Order No. 4348	2,532.82
Expenditure Order No. 5268	2,442.91
Expenditure Order No. 6682	2,151.36
Expenditure Order No. 7314 (March 1911)	1,993.39

Case 5: Fourteen-inch Disappearing Gun Carriage

Expenditure Order No. 4175 (January 1908)	$25,476.84*
Expenditure Order No. 7865 (October 1911)	17,981.44

* These figures include the cost of patterns, for which no separate cost figure is available.

The limitations of these figures are obvious. Some reductions in cost were to be expected as later orders were manufactured, if only because the workmen and foremen became more familiar with the job. The absence of a separate figure for the cost of patterns, which in three of the four instances distorts the first figure in the series, is not a defect of the historical record but an indication of an imperfect cost-accounting system before 1908. Before Wheeler instituted his reforms, no separate cost figures were recorded for patterns, and not even the arsenal officers knew what these costs were. It is unlikely, however, that in many instances the cost of patterns would exceed $1500.

Below are the remaining examples of reductions in total costs of production.[17] These are for the most part less detailed than the example already analyzed. Omitted from the list are cases in which there was claimed a reduction in actual costs below *estimated* costs; these seem hardly worth considering without more exact information as to how the estimates were made.

Case 6: Six-inch Disappearing Gun Carriages

	Under old system (1)	Under Taylor system (2)	Absolute reduction	Reduction as percentage of (1)
Labor costs	$10,229	$ 6,590	$3,639	35.58
Shop expense costs	10,263	8,956	1,307	12.74
Material costs	4,114	2,197	1,917	46.60
Total	$24,606*	$17,743	$6,863	27.89

* This figure includes the cost of patterns, for which no separate estimate is available.

Case 7: Castor Wheels for Shop Trucks (molding cost only)

Total	$ 1.45	$ 0.93	$ 0.52	35.86

Case 8: Elevating Arms for Six-inch Disappearing Gun Carriage (molding costs only)

Total	$42.35	$24.87	$17.48	41.28

Case 9: Large Wheels for Shop Trucks (machining costs only)

Total	$ 7.00	$ 3.67	$ 3.33	47.87

Case 10: Drain Plugs for Recoil Cylinders (machining costs only)

Total	$ 0.25	$ 0.07	$ 0.18	72.00

Case 11: Cylinder Plugs for Recoil Cylinders (machining costs only)

Total	$ 0.61	$ 0.09	$ 0.52	85.25

These statistics are inadequate to serve as the foundation for any generalized conclusions as to the effect of the Taylor system on costs of production. The sample is too small and may be biased; the breakdown of costs is, even in the best cases, insufficiently detailed; and no information is provided in most instances as to the size of the various orders for which cost figures are quoted. One would like to know, for example, whether the figure of $42.35 cited as the cost of production before the installation of the Taylor system in Case 8 relates to a single prototype or to a sizable batch. The difficulty is that the arsenal manufactured in quantity no standard homogeneous product which would enable us to estimate changes in cost over the period. Even for the varied products about which we do have information, the evidence typically consists of cost figures "before" and "after" the introduction of the Taylor system. But before and after can mean before and after the men involved were put on the premium system, or before and after the introduction of new costing methods, or before and after Barth had finished respeeding the machine tools. There is not even any sound basis for supposing that product specifications remained unchanged, except the questionable presumption that such changes would have been mentioned in the testimony if they had occurred. There is, in brief, no reliable evidence to show that the examples given are representative of a general process of cost reduction. With reference to Cases 9, 10, and 11, General Crozier stated that they were "not exactly typical" but rather the best out of the examples he had seen.[18]

The situation appears to call for hypotheses rather than firmly substantiated findings. Bearing in mind all the relevant considerations already spelled out with reference to Case 1, there is nothing incredible or even implausible about cost reductions of the magnitude claimed. Cases 9, 10, and 11, for which the largest percentage reductions are claimed, relate to small machined parts. The reductions in cost in these cases can be adequately explained in terms of the respeeding of machine tools and other strictly mechanical improvements carried out in the Watertown Arsenal machine shop. If, as a result of the use of high-speed steel, one of the drain plugs of Case 10 could be machined in half the time formerly required, a reduction in labor and shop expense costs of 72 per cent is entirely possible, apart from any bookkeeping reductions through new cost-accounting methods.

Cases 7 and 8 relate to the foundry only. Up to the time when these examples were quoted, the only parts of the Taylor system that had been applied to the foundry were time study and the premium system. The work of the foundry was not routed from the planning room and little if any systematization of working conditions had been done there. It would be of interest if separate estimates of the reduction in direct labor costs for these jobs were available, since it would enable us to assess the effects of the incentive-payments system separately from the systematization and standardization aspects of the Taylor system. Unfortunately, there are no such figures. In Case 7, however, the molder concerned earned a premium of 35.5 per cent over his day wage, and in Case 8 a premium of 42.5 per cent. This suggests that substantial time economies were achieved in these instances merely by the premium system.

Case 8 also illustrates the uncertainties which surround these figures. One of the molders employed in the foundry testified that this job was badly slighted after it was put under the premium system and that refinements, such as the shrinking brackets on the hub, were omitted; he also claimed that special facilities and assistance were given to the molder engaged on the job.[19] What this witness asserted was that the cost figures before and after did not refer to the same job; both what was produced and how it was produced differed.

Case 6 is comparable with Case 1. Labor costs were reduced by about one third, material costs by almost one half, and shop expense costs by over one tenth. The economies claimed are of the same order of magnitude as those claimed for Case 1, with the difference that shop expense costs apparently show a sizable reduction. The "before" figure includes, however, the cost of patterns; if these are estimated at a conservative $1000 and deducted from the shop expense figure, shop expenses show the same moderate reduction as in Case 1.

There are, then, two examples provided in tolerable detail, each of which relates to the arsenal's principal product: gun carriages. If we took these as representative, we could conclude that the installation of the Taylor system of management at Watertown Arsenal made possible a reduction in average costs of production of between one quarter and two fifths; labor costs were reduced by between 35 and 40 per cent and material costs by approximately one half; and overhead charges per unit of product remained approximately unchanged. If, accepting one possible interpretation of the evidence, we were to rule out the savings in material costs on the ground that they were merely paper economies, the reduction in total costs per unit of product would appear to have been about 15 per cent. Greater accuracy, in view of the paucity of evidence, is not possible.

Whether or not we accept these two examples as giving some credible indication of the general order of cost reductions which the Taylor system made possible depends upon the weight which we attach to a mass of supporting evidence. It seems by no means improbable that the introduction of the new system of cost accounting, combined with the general tightening ,up of stores procedures, produced economies in material costs of the order claimed. If this is accepted, the question at issue comes down to the inherent probability of the 35 to 40 per cent reduction in direct labor costs, combined with the maintenance of shop expenses at approximately the same level. Part of this alleged reduction in labor costs was achieved by the reclassification of workers — charging certain wages to shop expenses which formerly had been charged to direct labor. But this can hardly have amounted to more than a small percentage and would tend to be offset by a rise in shop expenses.

The only factor that would produce this reduction in labor costs and at the same time keep shop expense costs per unit of output unchanged is a substantial increase in the rate of manufacture. The question revolves around the size of the time economies in production. Colonel Wheeler's estimate was that under the Taylor system work was done 2.64 times faster than under the old system. This figure was first quoted in 1911, on the basis of a relatively small sample of jobs done under the premium system in the machine shop. Wheeler quoted it again, though with less pretensions at accuracy to two decimal places, four years later, when, in his interview with R. F. Hoxie, he stated that "The increase in output has been one hundred fifty per cent. We are doing work two and one-half times as fast." [20] If this figure of a 150 per cent increase in the rate of output is accepted, then direct labor costs would fall about 40 per cent and, even though shop expenses in the aggregate rose, the shop expense costs of each job would not necessarily rise in the same proportion and might even fall.

The plausibility of the reductions in costs of production claimed for the Taylor system rests therefore upon two assertions: that new cost-accounting and storekeeping methods led to reductions in material costs of approximately 50 per cent, and that the rate of output increased by approximately 150 per cent. Writing to General Crozier in April 1909, before Barth went to Watertown, Frederick Taylor had claimed "without hesitation" that the introduction of the Taylor system at Watertown Arsenal would make it possible either to double the amount of work done in a given time with the same number of men or to produce the same amount of work with half the labor force.[21] If the surviving evidence can be accepted, it would appear that Taylor, for once, understated his case.

The officers of the Ordnance Department were aware that, to a skeptical inquirer, the available information on cost reductions was far from incontrovertible proof of the beneficial effects of the Taylor system. Crozier, writing to Taylor in 1912, bemoaned the difficulty of proving beyond doubt the efficacy of the reforms that had been instituted. "We have not the supreme test of improvement exhibited by a surplus available for division which is available in a private establishment as a proof of the value of new methods, and it is much more difficult to establish convincingly, with us, that we have actually made improvements. . . . I would like to find some general measure of efficiency of the fiscal year 1912 as compared with some other year; but nobody has been ingenious enough to hit upon it." [22] Wheeler, replying to a request from the editor of the *American Machinist*, had to admit that "comparisons of costs are difficult, for the reason that there is no repetition work at this arsenal." [23]

This did not imply that Crozier and his subordinates ever doubted that the reforms they had carried out constituted a marked improvement and that the work of the arsenal was being conducted much more efficiently than before. They might not be able to prove it to others, but they themselves were convinced. Practical evidence was provided by the decision to extend to the other manufacturing arsenals — particularly Frankford, Rock Island, and Watervliet — the methods installed at Watertown. Crozier had always insisted that the Watertown installation was experimental; the decision to generalize the techniques demonstrated that the experiment was regarded as a success. Crozier seems to have regarded the installation of the Taylor system in the manufacturing arsenals as the culmination of his career. When in the fall of 1912 he gave up his position as Chief of Ordnance to accept what later turned out to be a temporary assignment as President of the Army War College, he reviewed in a letter to Taylor what he felt to be the concrete improvements he had made during his eleven-year tenure of office: "placing the personnel of the officers on a merit system, the adoption of wire-wound cannon, the establishment of courses of theoretical instruction and practical shop work for the officers, the improvement of accounting, and finally, the introduction of the Taylor system of scientific management." [24] Having done these things, he felt that it was time for a new man "with fresh ideas in the back of his head" to take over. "As for your system of scientific management," he went on, "I think it is in the Department to stay."

Had it not been for the acute public controversy stirred up by the use of the Taylor system, the officers of the Ordnance Department would have been satisfied with the success of their innovation. From their point of view it had proved its worth: the stated objectives of the installation had been largely achieved. From other points of view the situation was less satisfactory. The unions remained opposed, creating an ever-present threat of hostile political action. More immediately, the employees at Watertown refused to become reconciled to the system, finding in its day-to-day operation specific reasons for criticism and resentment.

What were the grounds on which the employees at Watertown Arsenal — the men whom Congressman Redfield of the House Committee on the Taylor System, a man not given to superlatives, called "an exceptional set of mechanics" [25] — objected to the Taylor system? Were their points of criticism specific or merely generalized expressions of dislike? Were there any significant differences in attitude between foremen and ordinary employees, or was the reaction much the same on the part of all concerned? Did they object to all aspects of the Taylor system, or only to some features?

There is ample evidence to answer these questions, particularly in the direct testimony of the employees at the hearings of the House Committee on the Taylor System and at Lieutenant Colonel Thompson's inquiry into the molders' strike. Other evidence is available — in letters of the employees to members of Congress and in the various petitions that were submitted to General Crozier and the Secretary of War after the strike — but it is colored by official union opinion and hence may not to the same extent reflect the personal views of the men. Whenever possible, therefore, the following analysis is confined to the personal statements of the men employed at the arsenal.

An episode related by Carl Barth may serve as introduction. Shortly after the molders' strike, Barth visited Watertown. Dwight Merrick persuaded him to meet Cooney, whose refusal to submit to time study had been the immediate cause of the strike, in order to see (as Barth put it) "what a fine man he is and what an exceedingly hard worker he is." Barth described his conversation with Cooney as follows:

I took great delight in making Mr. Cooney's acquaintance, and commented on the good reputation he enjoyed all around; and after a while I said: "Mr. Cooney, I guess by this time you realize that we are not as bad as you thought we were, and that our efforts are to help you instead of hurting you." He said, "Mr. Barth, our concern is not for the present. As things go now, here, nothing could be nicer; our concern is for the future." I said, "Mr. Cooney, I do not believe you need to have any anxiety for the future; at least we are working for the future more than for the present." He then told me that he could not altogether see through this scheme of ours, because he was not producing any more than he used to, and still he was getting a great deal more money. I said, "Mr. Cooney, you are mistaken. You would not get any more money unless you were producing more. I am afraid that you mistake work for production, and look upon your production merely as the effort you put forth." [26]

Cooney's remarks emphasize one theme that runs through all the testimony given by the employees: distrust. The men objected not so much to what was being done at the time, but to what they thought would be done in the future. They interpreted the meaning of immediate events in terms of their expected consequences and reacted to the imputed meaning rather than to the events themselves. The management of the arsenal were determined to make changes which they believed would be in the interests not only of the government but also of the employees. The men, experiencing these changes, refused to accept the assurance of ultimate benefit and foresaw instead consequences that they considered repugnant. Basic to the conflict was a distrust by one side of the predictions that were in sincerity made by the other.

To convince the employee that the changes made would benefit him directly and immediately, the Taylor system relied upon an incentive-payments system of one variety or another. It was essential that the men begin to earn higher pay as quickly as possible. The fact, rather than the prospect, of increased earnings was expected to overcome suspicion. The connection between increased pay and a reduction of distrust, however, was by no means inevitable, as Cooney's remarks demonstrate. Cooney's suspicions of the new system were increased by the fact of his higher earnings. He did not feel that he had earned the extra money. He was convinced that he was not producing any more. If, then, he was being paid more, there had to be some sinister purpose behind it. He did not see through the scheme and for that reason suspected it more.

Cooney was not alone in this. The general attitude of the men to the premium system was that there was a catch in it somewhere, that they were being bribed or fooled into doing something that was not in their interests, and that the higher rates of pay would prove only temporary. In 1912 General Crozier was disconcerted to be told that "the Watertown men consider the premiums as nothing more than gifts, claiming that they receive them without doing any more work." [27] A premium conceived of as a gift was not likely to serve as an effective incentive, nor would it lessen distrust. It must not be assumed that the men were being deliberately obtuse in not recognizing that productivity had increased. Willard Barker, a foreman in the machine shop, commented after the Taylor system had been in operation for two years that he had observed no great difference in the amount of work produced or in the speed of getting it out.[28] Olaf Nelson, master mechanic and head of the planning room, testified: "I dare say I do not get so much work in the shop per man today as I did two years ago, because they do not care if they hold their jobs or not."[29] The increases in productivity that were clear to the officers were by no means obvious on the floor of the shop. As long as this situation persisted, the higher earnings that resulted from the premium system injured rather than improved the morale of the employees.

Experience of rate cutting in other jobs was reflected in a widespread conviction that the higher daily earnings possible under the Taylor system were temporary. The men were given the usual assurances that a rate, once set, would never be cut unless the job was changed. Few of them seem to have believed this. Most realized that there were more ways than one to cut a rate. "It looks good on the face of it," said one molder, "to give me six hours extra pay when I have completed the job in 24 hours before. . . . That's pulling me right into the halter. . . . That is giving me money for nothing, and I have been fighting here two years for a quarter. But you can't shove it down my throat that that's going to last."[30] Orrin Cheney, a machinist, cited a case that was to him a flagrant example of rate cutting. A machinist's helper, earning $1.84 a day, was given a job that previously had been done by a machinist earning $2.66 a day, and the latter was transferred to a job previously done by a man earning $3.04 a day. Neither the helper nor the machinist received an increase in pay. Nobody's pay was reduced; but the rate for the job was cut.[31] Joseph Hicklin, a molder of thirty-eight years' experience, was opposed to the Taylor system "for the simple reason that I have worked piecework in jobs in the city of Boston." [32] James A. Reagan, a machinist, thought that the effect of the premium system would be to prevent the workman from getting raises in his base rate of pay and was convinced that the time on his job had been changed, despite assurances from Wheeler and Merrick that this would never happen.[33] Others believed that persistent failure to earn premiums would result in a reduction in efficiency ratings and therefore in base rates.

In general, and whether or not they felt they had to work harder or resented the pace set, the men displayed no confidence in the stability of rates under the Taylor system. They may have had some justification, for the officers themselves seem to have been far from clear as to what the promise not to cut rates entailed. Wheeler and Williams emphasized that they retained the right to change the rate if the methods of production were changed; this proviso opened the door to many exceptions. But Williams went further than this. Commenting in 1913 on a petition from three hundred and forty-nine employees at Watertown Arsenal which claimed, among other things, that rates set by time study had been cut, Williams asserted that "the rate set on any individual piece has never been changed" but added the damaging qualification that if the rate on a job was believed to have been set too high, management reserved the right to alter it when a similar job came up again. "This," he said, "is nothing more than making use of experience without which it would be impossible to run the shops." [34] By a rate that was "too high" Williams apparently meant one that enabled a man to earn more than a one-third premium. His position is understandable, but no more so than the position of a workman who could see no significant difference between this and the usual forms of rate cutting.

Taylor doctrine insisted that piece rates would never be cut because, if they were set by means of scientific time study, it would never be necessary to cut them. To some extent the men's lack of confidence in the fixity of piece rates at Watertown was founded on their distrust of time study, which took two forms: objections that it violated self-respect, and objections that it had not been carried out rigorously, and perhaps could not be, at Watertown.

Interesting examples of the first type of objection are to be found in the men's assertions that the work pace set by time study made it impossible for them to do the work as conscientiously as they wished. A feeling of pride in the standards and traditions of the craft is implicit in statements such as that of E. M. Burns, a Watertown machinist, who testified that "if I am given time on day work I endeavor to do that work to the very best of my ability; now, if I am given a bonus or premium to get out more work it is only natural to suppose that I would slight my work every way I can, just to get it by the inspection, in order to make more money. . . . It would hurt my reputation very severely, I think." [35] Alexander Crawford, foreman of the erecting shop, believed that the premium system led to "a tendency to slight the work . . . because a man naturally wants to make all he can";[36] and Isaac Goostray, a molder, believed that the long-run effect would be to reduce standards of craftsmanship. "It will make an inferior class of workmen, for the simple reason is that when a man is speeded up too much he will slight his work, and the consequence will be in a short time from now that if this system is carried on we will have an inferior class of workmen. The manufacturers at this time are seeking first-class workmen, and they cannot get hold of them, and this is a system that will make more imperfect workmen." [37] Particularly significant is evidence indicating a higher level of tension and insecurity among the workmen. The testimony of James A. Mackean, foreman of the south wing of the machine shop, illustrates the point. "The men," he stated, "at the present time are frequently punished for making mistakes, and sometimes they deserve it, but as foreman of the room it seems to me that the more often the men are punished for making mistakes the more terror-stricken they are. In my room at the present time I can scarcely walk down the room but what a hand is grabbed out here and a hand grabbed out there; and they seem fear-stricken that they are not competent to read the drawings correctly, or do the work correctly. It imposes a great deal of responsibility on me and it makes my work harder." [38] Mackean was skeptical about the ability of a time-study man, who "only has the business in theory," to suggest improvements in work methods: "They can give us certain speeds and feeds in figures; but they can not show us how to do the work better by any system than we know now. They can show us how to do it faster, but not better." [39] Olaf Nelson stated his belief that the determination of speeds and feeds should be left to the judgment of the mechanic and not prescribed by someone in authority. [40] If the machinist did not possess the knowledge and skills required for his work, he should be discharged.

Equally common were objections to time study based not on the damage it might cause to standards of workmanship but on the indignity that it was felt to involve. Sometimes this took the form of criticisms of the time-study man. Cooney argued that the man taking the time study should be "a practical molder . . . or at least a practical man"; the molders knew that Merrick had no foundry experience — "we knew when he started in to do this from the mere fact that he did not first of all have the pattern properly fixed, or the proper size flask" — and found it easy to fool him. "We understood from him that he intended to instruct us as to unnecessary labor that was put on the jobs. Our claim is that, not being a capable man, he is not able to do that." [41] In general the time-study man's qualifications were a secondary issue; the principal objection was to the fact of being timed. "A man does not mind how much the foreman watches him," said one Watertown molder, "or how much an officer watches him, but a man with a stop watch at your back all day long is an awful strain on anybody, and especially when that man is asking you all kinds of questions about the job." [42] Joseph Hicklin objected because "you are constantly being watched. There are men standing over you all the time and of course you are almost drove to it. You have got to keep pegging at it and working." [43] Cooney argued that the bias of the time-study man was inevitably in the direction of underestimation of the time: "He has at stake a system, and if he can take out of me a little more energy than what I have been putting out it adds to the good name of his system." [44] But more typical was resentment at being timed, regardless of any upward or downward bias. "I don't like a man to stand over me with a stop watch," stated one of the molders, "because it looks to me as if it is getting down to slavery to have a man following you when you are at your job; follow you when you are going to the dump, and with a stop watch stand over you while you bend down to pick up a few rods. That is why I call it so. This is too much for a man to stand." [45] Time study was, in short, regarded as a personal indignity, particularly by men who thought of themselves as experienced craftsmen.

The pace set by time study was frequently regarded as excessive. The men felt hustled and driven and resented the idea of being set in competition with each other. Asked whether working under the Taylor system had had any physical effect on him, Hicklin replied: "A great strain on my nerves . . . by the time I get through I am not able to do anything else when I get home. I am just fit for bed." [46] Other employees agreed with him; they went home more tired than when working for a straight daily wage. Considerable ill feeling was caused by a remark made by Dwight Merrick, who, when a machinist complained that he could not possibly equal the time set, told him — no doubt facetiously — that he would just have to run around the machine, when making his changes, instead of walking. The humor was not appreciated by men who regarded time study as implicit criticism of their work and an encroachment upon rights of personal privacy.

There is no evidence that the times set, either in the machine shop or in the foundry, were unreasonably strict. Any attempt to break through the comfortable and relatively leisurely pace of work that had established itself at Watertown would have caused resentment. The arbitrary manner in which the attempt was made and the failure to elicit active cooperation from the men aggravated this feeling. The atmosphere of conflict, uncertainty, and tension would be sufficient to create fatigue and send the men home exhausted.

In two cases, time study was criticized on the grounds that it led to a reduction, not an increase, in the rate of work. Nelson, whose testimony on the point is probably reliable, reported the case of a machinist whose normal rate of production was sixteen pieces an hour; after a time study had been made of the job and the machine respeeded and the man put on premium, the rate of production fell to twelve pieces an hour, and the man claimed that he was working far harder than he had before.[47] In the foundry, Gustave Lawson, when given his first job under the premium system, told his foreman that the time given him was much too long and that he would prefer to stay on a flat daily rate.[48] At Watertown cases of this type were exceptional. The men's distrust of the Taylor system never deteriorated to the point later reached at Watervliet Arsenal, where the foremen testified, in response to a request for information from their commanding officer, that the only reason the men and foremen did not like the system was because they could not work as fast with the Taylor system as they could without it. "The general

opinion of the men in the shops," stated the foremen at Watervliet, "is that this system was put in these shops to make
the work cost so that the private concerns would eventually
get the work away from this Arsenal . . . every man in the
shop knows the work is costing twice as much as it did under
the old way and they are naturally discouraged." [49] No such
fantastic misinterpretation of the motives of the Ordnance
Department developed at Watertown; the apparently haphazard results of time study, however, undoubtedly aggravated the men's distrust of the Taylor system and under-
mined their confidence in the management.

Principally in the foundry, but also to some extent in the
machine shop, there were allegations that the conditions of
work made accurate time study and equitable premiums impossible. Essentially these amounted to assertions that there
had been insufficient standardization. The molders described
in vivid terms the irregularities that made standard times and
bonuses unrealistic. "The shop is not adapted for a stop-watch
system," claimed Hicklin, "because there are so many delays.
You are working on the floor one day and moved off it onto
another [job] another day. Your helper is taken away and a
green helper sent you. You are waiting for clamps, gaggers,
etc., and have to go and find them yourself." [50] Describing
the foundry at Watertown as a "jobbing shop," Lawson
claimed that conditions made the Taylor system impractical:
"You work on one piece one day, and you work on another
piece the next day. You are shifted from one end of the shop
to the other; you have no steady place to work; you have no
practical helper — that is, one who understands you. Your
helper is changed on every mold you ram. Some molders get
a good helper and others a poor helper; whereas under the
premium system all molders ought to have equally good helpers in order to work under the same conditions, and I think
that this is impossible in a jobbing shop." [51] The annoyances

and petty injustices caused by lack of standardization were described by molder Martin Roach:

This morning, the first morning I have been put on premium, me and Hall were put on one job. We got the sand pile just after casting last night and it was red hot. That affects our molds, because it dries and sticks to the pan. We only had one helper for the two, and I asked Mr. Larkin for a helper, and he said that he only had nine helpers in the shop; four were outside. I said that I could not work very well without a helper. "Well," he said, "do the best you can, I cannot do anything for you"; so I went down the shop and spoke to Major Williams. Major Williams said that conditions were bad and to go to work, and I said to Mr. Larkin, "I am on day-work," and he said, "Who says so?" I said, "Major Williams says so." He said that I was allowed one hour, and I do not think one hour is enough under the conditions. If the sand pile is hot and the pattern close to the flask there is more finishing to be done.[52]

Thomas Kane, another of the molders, asked what were his objections to the premium system, replied simply, "Well, that we are not equipped. The principal objection to it is that we need equipment in there in the foundry." [53] Scarcity of mechanical equipment, when combined with the premium system, introduced an element of competition that the men resented; Kane particularly objected to the practice of taking the crane away from a man working by the day and giving it to a man working on the premium system. Earlier the use of the crane had been regulated by workshop custom; as Hicklin put it, "everyone has been willing to assist each other in giving up the crane." [54] Inadequate mechanical equipment made it impossible to standardize working conditions and aggravated the disruptive competition implicit in the premium system.

Very little systematization was done in the foundry before the introduction of time study. Complaints that time study and an equitable premium system were impossible in such conditions were justified, as Taylor would have admitted. In the machine shop, which had undergone considerable preliminary reorganization, such complaints were less common. They were not, however, nonexistent. Richard H. Stackhouse, a machinist, asserted that the foremen frequently changed speeds and feeds on their own authority, regardless of what was prescribed on the job card.[55] One of the foremen in the machine shop, Mackean, testified that there had been "considerable destruction of countershafts and shafting, due to tight belts and high speed," which inevitably caused delays and interruptions,[56] and further that the routing system — "a splendid thing if carried out" — had worked imperfectly: "one-half of my duties is to hunt work for the men." [57] He also complained of the lax way in which some times had been set. In one case steam pressure in the boilers had been very low, and in addition, "Mr. Merrick came out and stood at that machine and talked three-quarters of an hour"; naturally, the job had taken much longer than was really necessary, and Major Williams had taken him and the machinist to task for it. "As the shepherd over the flock there I must object, and I must give the men justice as well as the system justice." [58] Olaf Nelson was more inclined to complain that the system as a whole was too inflexible; tools ground to Taylor specifications were all right for roughing work but not for finishing, and 50 per cent of the work at the arsenal was finishing. Nelson reported that the men objected because they were not permitted to grind their own tools for finishing, a complaint with which he was evidently in sympathy.[59]

Unqualified condemnations of the entire Taylor system are to be found neither in the testimony of the workmen nor in that of the foremen. The men objected to features that they found objectionable, such as the indignity seen in time study, the accelerated work pace, and the disruption of traditional pay differentials. The foremen were more inclined to question whether there was any net advantage to be derived from the new system, balancing the unfortunate effects on morale against any possible increases in the rate of output. On incentive payments they were explicit: a straight daily wage was better than piecework, and straight piecework was better than a premium system. "I do not think the premium system of wages is really the proper thing," said one of the machine-shop foremen. "I am myself more in favor of the straight day wage, and the men being compelled to do a good, square day's work for a good, square day's pay." [60] "If you want piecework," advised Nelson, "have straight piecework and not stop-watch systems . . . the one bad part of the system is this stop-watch or premium system; nothing else." [61] To the suggestion that incentive payments were necessary to induce the men to put forth their best efforts, foreman Mackean had a simple answer: "If a man is so lazy — to use no better word — that he will not do a day's work without being put on a premium system, he should be immediately removed from the shop and a better man put in his place." [62] Personally he was unconvinced that the premium system had brought with it any increase in the rate of production: "Before this time Mr. Nelson was speeding everything to about the top notch as foreman, before this system came in, and I do not see, though, that we are producing much faster, but, of course, the figures will settle whether we are or not. But personally we do not see, that is, the foremen and the workmen, where there has been a great deal more produced, but from the figures that we have heard from the officers it appears so." [63]

In all this testimony it is difficult indeed to see any direct reflections of union propaganda. The criticisms made of the Taylor system by workmen and foremen were for the most part specific and pointed. They had no need of any outside organization to put words into their mouths. They knew what they disliked, what seemed to them unfair, and what, in the circumstances, was an unsound way of getting the work done. Seldom in the testimony is there any hint of organizational pressure brought to bear upon the employees by their unions. The objections expressed by the men were reasonable and well founded. They spoke of things they knew, and spoke forcefully for that reason.

Apart from the congressional hearings from which most of the above testimony has been drawn, two attempts were made, in the period between the molders' strike of 1911 and the congressional prohibition of time study in 1915, to survey the opinions of the Watertown employees as a group. One was conducted by the officers of Watertown Arsenal on orders from General Crozier in August 1913, the occasion being a mass petition from the Watertown employees to the Secretary of War against the continuation of the Taylor system. Crozier was disturbed by the petition and its possible political repercussions; he ordered the inquiry in order to be reassured that the men, interviewed individually, were really not opposed to the system. The survey was partial, only forty-five out of the three hundred and forty-nine men who had signed the petition being questioned; it was formal, essentially a cross-examination of the men by their superior officers; and it was abruptly halted when the men complained to Congress that they were being intimidated. The findings were not very informative. Major Williams, summarizing his major conclusions, reported that "the attitude of the men at this arsenal is believed to vary exactly with their Union affiliations. Those who feel very strongly in sympathy with Unionism are equally strong against the Taylor system and all other systems which depart from the straight day wage. The weaker this feeling of sympathy becomes, the less opposition there is to the Taylor and similar systems." [64] Williams does not appear to have considered any variable other than sympathy with unionism.

The second survey was more ambitious and more enlightening in its conclusions. Early in 1913 an independent management consultant named Miner Chipman was retained by certain employees at Watertown Arsenal, acting through an informal "Workers' Committee" to make a study of their complaints and prepare a brief on their behalf to be submitted to the Secretary of War.[65] A former associate of Harrington Emerson and an early member of the Society for the Promotion of the Science of Management (later the Taylor Society), Chipman was by no means unknown in management circles, though not one of Taylor's personal coterie. Taylor heartily disapproved of Chipman's inquiry and did his best to discount its conclusions. "Of course," he wrote to Barth, "you appreciate that Chipman is in this thing for the money there is in it — nothing else — and I do not think he will cut much ice one way or the other." [66] To General Crozier he offered much-needed reassurance: "We all know that the system is splendidly applied at the Watertown Arsenal, in spite of what he [Chipman] says; also that the workmen who are working under it are mighty well contented and happy." [67] Crozier, however, was not to be diverted; if Chipman had information, he wanted it. In January 1915 he had denied Chipman access to Ordnance Department records; December of the same year found him respectfully requesting from Chipman any guidance he could provide as to "the real feelings" of the men at Watertown.[68]

Chipman spent eleven months on the investigation and ended it, by his own account, $2000 out of pocket. He later asserted that his work had been "looked upon with suspicion by the officials of organized labor," and there is no evidence that he received any assistance or cooperation from the unions. His findings were based partly on conversations with the men at Watertown, but principally on an elaborate questionnaire that he mailed to the home address of each employee. The response was not unsatisfactory: of 342 blank questionnaires sent out, 235, or 68.7 per cent, were completed and returned.

A copy of the questionnaire is not available, and we must rely on Chipman's summary of the replies. The intent of certain of the questions is not very clear. Chipman apparently thought it worth recording, for example, that the married men who opposed the Taylor system had an average of 2.9 children each, while the married men who did not oppose it had an average of 4.4. Some of his findings are more to the point. Of the 235 men who replied to his questionnaire, 214 stated that they were opposed to the Taylor system. Noting that the Ordnance Department had claimed for years that it was the union men at the arsenal who were opposed to and had agitated against it, Chipman reported that, according to his information, a majority of those opposed to the system were not union members. Of the 214 respondents who stated that they were opposed, 113, or 52.8 per cent, were nonunion men and 101, or 47.2 per cent, were members of unions. Asked "Do you think the agitation is brought about through Union labor or similar sources?" 137 answered "No," 28 "Yes," and 70 ignored the question.

The statistical results of Chipman's questionnaire are open to question. No attempt was made to follow up those employees who did not answer the questionnaire, and for those who did answer there was no internal check on consistency. Some of the statistical differences that Chipman apparently thought important are of dubious significance. What, if anything, does it imply that the average daily wage of those opposed was 21 cents higher than that of those not opposed? Or that a slightly higher proportion (96.2 per cent) of the single men were opposed than of the married men (90 per cent)? The usefulness of the replies was severely limited by the crude frame of reference used in the questionnaire. A respondent had to be opposed or not opposed, or he was not counted. The Taylor system was presented as a single undifferentiated thing, with no provision for the possibility that some aspects might be liked but others disliked. Crude categories of this sort inevitably did violence to the true complexity of the situation.

Chipman's abilities showed to better advantage when he moved to the less confining tasks of commentary and criticism. In his report on the Watertown survey to the Efficiency Society, he showed considerable insight into the causes of conflict at the arsenal and presented an analysis that moved on a different intellectual plane from any examined up to this point. Scientific management, he asserted, as Taylor had presented it and as it had been applied in industry, lacked three elements of fundamental importance: an adequate method of industrial education, an adequate recognition of the principles of democracy in industry, and an adequate conception of, and sympathy with, the sociological and economic aspirations of the worker. Education — to accept and adapt to new forms of organization and new methods of work — was specified as a critical necessity. "The breaking-up of crafts under scientific management," said Chipman, "is also breaking up certain of our social formations. Industrial training must be more than the mere training for a particular piece of intensive production work. It must give the worker an adequate conception of his place in the whole process of manufacture." [69] Conflict was inevitable, he argued, if change was forced on men regardless of their wishes. The course of innovation could not be smooth unless adequate consideration were given to the worker as a human being and a serious attempt made to enlist his voluntary cooperation. Here was foreshadowed the emphasis on consent that was, after Taylor's death, to become the revisionist doctrine of the progressive elements in the Taylor Society.

Speaking of the situation at Watertown, Chipman denied that the fundamental ground of complaint was time study and the premium system. These were the aspects of the Taylor system upon which the objections of the workers had seemed to center. But to stop at this level of analysis was to concentrate on symptoms only. Not in the methods — whether the stop watch or differential pay — but in the way in which they had been introduced and administered was the root cause of conflict to be found. From the statements of over two hundred employees that he had examined, said Chipman, "it was made clear that the objection to these devices of scientific management was based upon the method of introduction and administration, and not upon the devices themselves." There had been "an absolute lack of the democratic spirit. . . . Stop watch and premium were taken as the targets of attack because they were considered to be the brain and heart of the system. Kill them and the thing would die." But was it inevitable that time study and incentive payments should cause this reaction? By no means, argued Chipman. His experience indicated that workers, even union members, had no objection to time study "where principles of industrial democracy are taken into consideration, and an educative process developed to meet the requirements of the situation." There had been none of this at Watertown. "The fight against scientific management at Watertown Arsenal," he concluded, "is, in reality, a fight against its arbitrary and dictatorial approach to the human problem. . . . Organized labor has taken a militant stand against the system as a whole because it has not taken into consideration principles and formulas of human welfare as well as principles and formulas of productive efficiency."

To remedy the conflict that had developed at Watertown and to provide a model for future installations of the Taylor system, Chipman proposed a system of committees to insure participation by the employees in the determination of piece rates and standard times. He laid down as a principle that "no time study shall be accepted as the basis for standard time or piece-work schedules, until the same has been approved by the worker, or a committee of workers" and recommended that a committee should be elected in each department of the arsenal to cooperate with management in the taking of time studies. This would presumably become a permanent feature of every department in which scientific management had been applied. In addition, he recommended that committees should be selected by the employees in each of the government arsenals, by the officers in charge of those arsenals, and by the American Federation of Labor, to meet in conference and, with the aid of expert counsel, draw up and submit to the Secretary of War and to Congress a set of proposals that would permit scientific management in government work to function on a permanent basis.

Chipman's analysis of the sources of conflict at Watertown showed more insight than those offered either by Taylor and his associates or by the officers of the Ordnance Department. His suggested remedies, however, though they were to have an important influence on future policy in the Ordnance Department, were offered too late to reverse the tide of union opposition. If the only obstacle had been the grievances of the men employed at the arsenals, Crozier's prediction that the Taylor system was "in the Department to stay" might have proved correct. Crozier refused to believe that the arsenal workers were really as hostile as the unions alleged, but he was prepared to go out of his way to satisfy any grievances that he knew to exist. Had he not as early as 1911 conceded that he could see no reason why all the "systematization procedures" of the system could not be introduced without time study and incentive payments? [70] Had he not admitted that the base rate of pay could be the subject of collective bargaining, and perhaps even the premium rate as well? [71] Perhaps Crozier was not yet ready to implement these ideas, but the seeds of later innovations at Watertown had been sown.

Such measures of reconciliation were not immediately feasible. The union's campaign to outlaw time study by political action had gone too far to be halted by piecemeal reforms. Ever since, at Rock Island Arsenal, Hobbs had made the first fumbling attempts to install time study and premium payments, there had been imminent danger of political intervention. Organized labor grew more and more determined in its opposition as time went on, until the elimination of the Taylor system in government work became a primary objective of union strategy. The strike at Watertown Arsenal provided the publicity and the emotional focus for an intensification of the campaign, and in the years after 1911 the possibility of a congressional prohibition of time studies and premium payments on work done under government contract was never far distant.

This was the reason for the apparent placidity of labor-management relations at the arsenals after 1911. The battle had moved to a different sector, one in which the decisive engagements were no longer to be won by strikes and petitions but by pressure on Congressmen and the threat of "the labor vote." Crozier was aware of this. From the beginning he had warned Taylor that the really serious threat to the experiment upon which they were both engaged was a political one. After 1911 the fate of the Taylor system in the Ordnance Department depended upon Crozier's political adroitness in Washington. And in the end, skilled though he was in the ways of politics, his abilities proved inadequate.

Crozier's strategy had two main elements. First, he wanted a complete and impartial investigation of what the Ordnance Department had done to install the Taylor system at Watertown and what the results of this policy had been. This, he sincerely believed, would result in the complete vindication of the system and of the department. Second, he wanted to block the passage of any hostile bills or resolutions that might be submitted to Congress. If such legislation came to a vote on the floor of the House, he was convinced that it would pass. His policy was to see that such resolutions were referred to committees that he could influence, where they could die quietly; as a last resort, he counted on the Senate to prevent their passage into law.

In his first objective he achieved only partial success. The House Committee on the Taylor and Other Systems of Management, authorized in the summer of 1911,[72] held prolonged hearings and took extensive testimony from the employees at the arsenals, from Taylor and his friends, and from the officers of the Ordnance Department. Its report, submitted on March 9, 1912, endorsed in full neither the department, nor Taylor, nor the unions, but merely stated that it was inadvisable to make any recommendations for legislation on the subject at that time.[73] This was far from the vindication that Crozier had hoped for, and it proved no stumbling block to Representative Pepper, who almost immediately introduced a bill to prohibit the use of the stop watch and premium or bonus systems in government establishments. A similar bill was introduced into the Senate. Both were killed by delays, but the respite was only temporary.[74] Another opportunity for thorough investigation of the Ordnance Department's policies did not arise until 1914, when the United States Commission on Industrial Relations began its inquiry; by the time this report was available, however, Congress had already acted.

In blocking or sidetracking proposals for legislative action against the Taylor system, Crozier's efforts paid more generous dividends. His policies were based on the assumption that the union agitation against the system had no foundation in fact; if congressional action could be postponed for a year or two, the probabilities were that all would be clear thereafter. This assumption was not entirely correct, and

Crozier seems to have underestimated the determination with which the A.F.L. and its affiliated unions were prepared to press for action.

He cannot be accused, however, of underestimating the probability that legislation condemning the Taylor system would receive congressional approval if it came to a vote. Taylor was inclined to interpret the real danger as possibly coming from the President and the Secretary of War. "The most important thing to do," he wrote to Barth in August 1911, "is to strengthen General Crozier in his relations with the Secretary of War. . . . It would, of course, be a great misfortune if the Secretary of War acting for Mr. Taft were to shut down on us." [75] And in July of the following year he advised Crozier to get as many as possible of the Watertown employees on premium "by hook or by crook" before the next administration took office.[76] Noting that Senator Borah — "one of Roosevelt's right-hand men" — had expressed himself as strongly in favor of anti-time-study legislation, he commented: "This gives us a pretty good sample of what would happen if Roosevelt got into the White House. I have not the slightest doubt that he would pitch scientific management to —— [sic] the very first time that he was asked to by the labor leaders." [77]

That Crozier had need of this amateurish counsel is doubtful. As an experienced administrator, he made sure that his relationships with his superiors were secure; throughout the whole period during which the Ordnance Department was under attack there were no signs of deterioration in his relations with the White House or the Secretary of War.[78] His relations with Congress, however, required careful maneuvering. By December 1912 resolutions were pending in both House and Senate for the suppression of time study and the premium system in the arsenals, and Crozier was exerting all the pressure he could on individual legislators to prevent

these resolutions coming up for consideration.[79] Successful in
that year, he found a more serious threat arising in the following one, when Representative Deitrick, from the Watertown district of Massachusetts, introduced a bill that would
have effectively nullified much if not all of the Taylor system
at the arsenals. If passed, the bill would have canceled the
wages of any officer or other supervisor having charge of the
work of an employee of the United States government who
was found making, or having made, time studies with a stop
watch or other measuring device; and it would have prohibited the use of any part of a federal appropriation for the
payment of a premium or bonus.[80] This threat too was temporarily neutralized, and the bill was never reported out of
committee.

By 1914 the situation had become critical, and it was evident that a showdown could not be long delayed. Taylor, impressed by an editorial supporting his system in the *New
York Evening Post,* urged Crozier to "get the press of the
country to take it up, either by paying them money or without doing so" and offered to supply some of the necessary
cash himself.[81] But Crozier preferred less chancy tactics:
"Any use of money . . . would be a mistake, and would lend
color to a charge that all of the newspaper comment should
be considered as tainted. Newspaper publicity is always
something which has to be handled with tongs." [82] His efforts
were now concentrated on Representative Deitrick, a potential source of trouble. Crozier understood Deitrick's position
but hoped that he could be made to see reason. "Deitrick is a
new Member of Congress," he wrote to Taylor, "who has
made a distinct bid for the labor vote. In a speech-making
tour of his district he made one of his speeches in an automobile directly in front of the arsenal gate at Watertown, and
in it promised to do all in his power to abolish the "Taylor
System" if elected. He is therefore, of course, taking steps to

make good on his promise. I do not think that personal conviction is a thing which he needs; but a realizing sense of how much hard merit he is up against." [83]

Crozier's persuasion was in vain. Deitrick, exposed to countervailing pressure from the unions, lived up to his campaign promises and in the next session of Congress reintroduced his proposals, but this time in a form more difficult to combat. The provisions of his earlier bill were now introduced as an amendment to the Army Appropriation Act, and in that form passed the House with only token opposition. A similar measure introduced as an amendment to the Navy Bill two weeks later also received prompt approval.[84]

Crozier's only hope now was that the Senate would reject the amendment and maintain that position in Senate-House conference. To strengthen his hand he resorted to tactics of doubtful expediency, and in January 1915 issued orders directing the immediate suspension of time studies and premium payments at all the arsenals. This paradoxical course of action can be understood only by reference to Crozier's unshaken conviction that the employees at the arsenals really approved of the Taylor system, no matter what the unions might say. His expectation was that the suspension of premium payments would (in his own words) generate such a "howl of opposition" at the arsenals that attempts to destroy the system would be discredited. Results of this last-ditch expedient were negligible, although a petition against the Deitrick amendment was signed (under duress, according to the unions) by several hundred employees at Frankford Arsenal. The Senate, after considerable debate, divided the amendment into two parts, one dealing with time study, the other with incentive payments. Despite vigorous support from a number of senators of widely different background and affiliation — notably Hughes of New Jersey, La Follette of Wisconsin, and Lodge of Massachusetts

— both sections were defeated. Crozier breathed freely once again. "I do not anticipate any danger of the legislation going back on the bill in conference," he wrote to Taylor.[85] He was wrong. The Senate-House conference took up first the amendment to the Navy Bill. The Navy Department, which was not at that time employing any of the practices prohibited in the bill, offered no objections; the Senate conferees yielded; and the amendment was approved. When the committee turned to the Army Bill, the precedent was allowed to stand, despite the known opposition of the Ordnance Department and the Secretary of War. As Crozier put it, analyzing the vote in retrospect, "The Senate conferees, having yielded on the Navy Bill, and the Senate having sustained their action, the House conferees on the Army Bill did not care to stand for what they would consider martyrdom for the sake of that bill alone." [86] If the Army Bill had been taken up first, the result might have been different.

And so, through the determined efforts of a well-organized pressure group and the vagaries of democratic politics, prohibitions designed to prevent the use of the Taylor system of management in government work acquired the status of law.[87] Had they been presented to Congress as a separate bill, they might have received a presidential veto; as a rider to the Army Bill, no such action was conceivable. It became possible to assert that the Taylor system had been repudiated by the American government and that, whatever private employers might choose to do, in government work the system had been weighed in the balance and found undesirable. And this is what organized labor did claim.

But did the bill prohibit the use of the Taylor system, as appeared on the surface? In two respects, one trivial and the other fundamental, it did not. Most of the work done at Watertown was financed from appropriations under the Fortification Bill, not the Army Bill. Since no similar restrictive amendment had been inserted in the Fortification

Bill, it remained possible to operate the premium system at Watertown, although time studies had to be suspended since the salaries of the time-study men came from Army Bill appropriations.[88] Crozier asserted, in answer to protests that he was violating the spirit if not the letter of the law, that Representative Deitrick had been aware of the fact that the provisions in the Army Bill would not affect the payment of premiums at Watertown and that he could easily have had the same restrictions added to the Fortification Bill if he had wanted to.[89] Deitrick's unexplained failure to insure that all the department's funds were restricted as were funds appropriated under the Army Bill meant that for one more year the premium system remained in operation at Watertown. The only change immediately necessary at that arsenal was the suspension of time studies. To anyone concerned to maintain the Taylor system, this was, of course, a serious limitation; temporarily it might be possible to set tasks on the basis of the recorded results of previous time studies, but in the long run, as jobs and equipment changed, task setting and the determination of piece rates would deteriorate into the haphazard business that the Taylor system had been designed to remedy. Crozier hoped that the prohibition of time studies was only temporary and that, by the time the next year's appropriations came up for approval, the furor would have died down. In this he was disappointed, for in the following year the restrictions were extended to cover all work done with federal government funds, whether under the Sundry Civil Appropriations Act, the Fortifications Act, or the Army and Navy Acts. With this demonstration of the intent of Congress, Crozier had no alternative but to submit.

For the next thirty-three years these restrictions were re-enacted annually, degenerating in time into an archaic feature of the appropriations bills whose origins nobody remembered and whose significance few understood. Presumably throughout this period, until the joint efforts of Senators Taft of Ohio and Flanders of Vermont had them removed from the statute book in August 1949,[90] the restrictions were enforced by the Ordnance Department.

Did this mean that the department, committed to the Taylor system in the years between 1909 and 1916, was compelled thereafter to remove the system from its operations and return to the obsolete managerial methods of an earlier age? The philosophy and practices of the Taylor system, once established, were not to be uprooted so easily. What had been prohibited was a particular method of wage payment and the use of the stop watch to set tasks. However controversial these particular techniques might be, whatever resentment they might cause, they were not the core of Taylorism. They were mechanisms only; they could be dispensed with and substitutes found if necessary. Straight piecework had not been declared illegal, nor had the use of time-study data obtained from similar jobs in other establishments, nor had the developing arts of motion study and job simplification. If such devices were essential, equivalents could be found that the law did not prohibit. Besides all this, there was left untouched by legislation the new attitude to management that the Taylor system implied — the idea of management as something that could be systematically studied, analyzed, and improved. Untouched too was the whole complex of organizational and procedural reforms that were required by the Taylor system in any establishment where, as at Watertown Arsenal, it had been thoroughly applied. The stimulation features of the Taylor system (time study and incentive payments) were only one element working in the direction of increased efficiency; in all probability the strictly mechanical and organizational changes made the major contribution to cost reduction. To argue that, once time study and incentive payments were prohibited, the entire Taylor system was frustrated is to fall into the fallacy that trapped many of Taylor's critics and, indeed, many of his friends. Did Barth's systematization of belt maintenance cease because Congress had prohibited time study? Did the machinists at Watertown once again find it necessary to forage for material and tools? Were the routing and scheduling of work from the planning room abandoned? Did cost accounting relapse into chaos? Obviously not. There was more to the Taylor system than time study and the premium system. The changes that in the long run were likely to make the greatest impact upon a plant's performance and growth were left intact.

If the installation of the Taylor system at Watertown Arsenal had included nothing but time study and premium payments, some slight reductions in costs might have been achieved at the price of damage to morale; but the improvement would have been short-lived. The important benefits that the arsenal derived from the Taylor system were of another nature, and their significance became apparent only in the long run. Fundamentally, the contribution of the Taylor system was to give the arsenal new potential for growth. An obsolete and restrictive organizational skeleton had been removed and a new one substituted. Watertown Arsenal, after the introduction of the Taylor system, was a healthier organization than it had been before, because it was an organization that could grow. Specifically, it was an organization that could, without damaging increases in costs, handle many times the volume of production that it had been capable of before. It is therefore in a sense of minor importance whether or not costs of production at the old level of output were reduced; the essential point is that new levels of output were now possible. What had formerly been the fixed factor of production — the organization itself — had now been made variable, and the specter of diminishing returns had once again been exorcised.[91] In the years after

1916, when the productive facilities of the Ordnance Department were stretched to the limit to meet the needs of World War I, this was to prove an accomplishment of no small value.

The other long-run repercussions of the installation of the Taylor system at Watertown, and of the controversy that resulted, are more speculative. Other writers have described the dramatic shift in the attitudes of organized labor toward productivity in the United States after World War I and the curious turn of events by which the Taylor Society, at one time the archetype of all that organized labor detested, became in the 1920's the foremost advocate of union-management cooperation.[92] It has not escaped attention that the designer of the famous Baltimore & Ohio plan, one of the first and most successful experiments in labor-management cooperation, was a former officer in the Ordnance Department, or that, in one pathbreaking instance after another, representatives of the Taylor tradition in the 1920's pioneered in the introduction of techniques of joint consultation in industry. It may not be entirely fanciful to see reflected in these later developments the lessons learned, by the unions and by the Taylor group, from the Watertown experience; nor is it entirely unrealistic to suggest that the leaders of the A.F.L. and its affiliated unions, presenting themselves after 1918 as supporters of any measures designed to increase industrial productivity, provided that they were introduced with union cooperation, were influenced by the education received in the prewar fight against Taylorism. These are causal connections not susceptible to proof, but it is hard to deny the existence of some relationship.

The persistent influence of the Taylor group upon the Ordnance Department, and indirectly by that channel upon American industry, is indisputable. After Taylor's death in 1915, as other writers have shown, the philosophy of the Taylor Society swung with remarkable rapidity away from the doctrinaire authoritarianism of the master toward a more comprehensive view of the principles that should govern the relationships of men at work. Typical of the new theories was Valentine's insistence upon consent as the indispensable prerequisite for all changes in working conditions; Chipman's arguments for labor participation in the determination of wages and tasks; and, above all, Robert Wolf's redefinition of efficiency in terms of the aspirations of both parties to the labor contract. "No matter how skilfully management determines the one best way to perform an operation," said Wolf, "it ceases to be the one best way if the workman does not want to do it that way." [93] This was the still radical point of view that, in the years after 1915, was gradually establishing itself among the heirs to the Taylor tradition.

And what of the Ordnance Department? That some of the officers in the department were thinking along parallel lines is evidenced by Crozier's willingness, as early as 1911, to admit that piece rates could be the subject of collective bargaining. War brought these two trends of thought together and transformed them into the official policy of the Ordnance Department. With the tremendous expansion of the Ordnance Department after 1917, experienced managers and executives prepared to enter that branch of federal government service were in short supply. Crozier and his subordinates filled the gaps in large measure by hiring members of the Taylor group. It has been estimated that by the beginning of 1918 one third of the members of the Taylor Society were working in the Ordnance Department.[94] These men took into government service the philosophy of industrial management that had germinated in the years of peace. In the Ordnance Department they found an organization that was not only favorably disposed toward this philosophy, but had already, in part, implemented it.

As early as 1915 Crozier demonstrated his ability to learn from the controversy created by the Taylor system. One of the complaints of the molders at Watertown was that they had no recognized method for voicing their grievances to the arsenal management. The strike had been precipitated by the breakdown of the informal procedures that had evolved to meet the need for communication between officers and men. Crozier's innovation — and there is no evidence that he acted under any compulsion other than his conviction that it was desirable — was the introduction at Watertown of a system of grievance committees, at different levels in the organization, on each of which both employees and management were represented. The influence of Miner Chipman's theories is clear. Colonel Wheeler, still in command of the arsenal in 1915, described these grievance procedures as "new in the Department, of a tentative nature, and necessarily . . . subject to the test of practical experience." [95] There was no provision for union recognition or for collective bargaining in the usual sense. Grievances were to be submitted to a shop board, consisting of one of the officers and the employees of the shop where the problem had arisen. If the grievance could not be adjusted at this level, it was submitted to the arsenal board, consisting of an officer of the arsenal selected by the commanding officer and a representative of the employees selected by them. Failing settlement at this level, the dispute was to be carried to a mediation board, which was to consist of five members: one officer appointed by the commanding officer to serve continuously on the board; one other member to be selected by the commanding officer, who might or might not be an officer but who was not to be a party to the dispute; two representatives chosen by the employees, one to serve continuously and one to represent the craft involved in the dispute; and a chairman selected jointly by the commanding officer and the employees. Provision was also made for the ultimate submission of the dispute to a three-member supreme mediation board, to include an officer of the Ordnance Department, a member appointed by the crafts representing the employees in all the arsenals, and a neutral chairman. The right of appeal to the Secretary of War was retained by the employees.

This hierarchy of boards was designed to rule out any possibility of a recurrence in the Ordnance Department of the kind of unpredicted conflict that had occurred at Watertown. The final clause of the agreement reserved to the employees the right to object to any manner of work at the arsenal "such for instance as the stop-watch elemental time studies or the premium system of payment" and reserved to the department the right to install such systems — an indication of its early inception. The agreement remained in force, however, after the congressional prohibition of time studies. It should be classed as one of the earliest examples of formal grievance procedures in American industry.

With the advent of war, similar procedures were adopted throughout the Ordnance Department. Ordnance Department General Order No. 13, dated November 15, 1917, and thought to be the handiwork of Morris L. Cooke, a leading member of the Taylor group, listed a number of recommendations to purchasing agents designed to insure that products manufactured for the department were produced under acceptable working conditions. Restriction of the workday to eight hours was recommended, along with limitations on the physical work required of women, and child labor was prohibited. Significantly, the order concluded by recognizing "the need of preserving and creating methods of joint negotiations between employers and groups of employees." [96] The Ordnance Department, in its own manufacturing plants, went further than this. At Frankford Arsenal to begin with, and by the end of the war at the other arsenals also, the employees were granted the right to choose their own foremen and to approve all piece rates in return for a promise not to restrict output. Nor was the trend reversed with the end of the war: the Arsenal Orders Section, established under the charge of Otto Beyer to bid on government manufacturing contracts and thus ease the arsenals in their transition to a peacetime basis, included on its executive board elected representatives of the employees at each arsenal as well as officers of the department. Joint responsibility and consultation were by this time, in the Ordnance Department, no longer an idea but a normal practice.

Both organized labor and the heirs of the Taylor tradition learned much from the enforced cooperation of the war years. The unions, in their postwar zeal for productivity, demonstrated their new awareness that, in a hostile climate of opinion, organized labor could not safely allow itself to be depicted as a massive deterrent to material progress. The representatives of the management movement showed by their insistence on consent and collective agreement that they had lost their earlier fear and distrust of unions and were prepared to take a broader view of the social dimensions of efficiency. The Ordnance Department was of course not the only organization in which this process of mutual education took place. The Shipbuilding Labor Adjustment Board, the Board of Railroad Wages and Working Conditions, and the National War Labor Board fulfilled, as a byproduct of their formal operations, a similar function. To the Ordnance Department, however, must be accorded the credit a pioneer deserves. Throughout the period with which this story has dealt, from the initial negotiations with Frederick Taylor in 1906 to the formal recognition of workers' committees in 1915, the department never stopped learning and innovating. The adoption of the Taylor system was merely one step in the process; the success of this step is to be measured, above all, by the degree to which it made further learning possible.

BIBLIOGRAPHICAL NOTE

The principal sources for the material in the preceding pages are The Taylor Papers, a collection of Taylor's personal correspondence and other memorials in the library of Stevens Institute of Technology, Hoboken, New Jersey; the records of the Ordnance Department, United States Army, which may be consulted at the National Archives, Washington, D.C.; and the testimony presented to the Special Committee of the House of Representatives to Investigate the Taylor and Other Systems of Shop Management, under authority of House Resolution 90 (published in three volumes in 1912).

Other important contemporary sources consulted were the testimony taken by the Interstate Commerce Commission in 1910–1911 during its investigation of certain proposed advances in freight rates by carriers in official classification territory; the Annual Reports of the Chief of Ordnance; the *Final Report and Testimony* of the U.S. Industrial Relations Commission of 1912; and the files of the *American Federationist*, the *International Molders' Journal*, and the *American Machinist*. There is also on deposit in the Library of Congress a collection of the papers and correspondence of John P. Frey, but unfortunately an opportunity to consult this collection did not arise during the preparation of the manuscript.

For secondary works, the interested reader is particularly advised to consult Jean T. McKelvey's *AFL Attitudes toward Production, 1900–1932* (Ithaca, N.Y., 1952) and Milton J. Nadworny's *Scientific Management and the Unions, 1900–1932: A Historical Analysis* (Cambridge, Mass., 1955). Also relevant are George Filipetti's *Industrial Management in Transition* (Chicago, 1946) and the special issue of the *American Journal of Sociology*, edited by Everett C. Hughes in March 1952 (vol. LVII, no. 5), devoted to the sociology of work.

A comprehensive bibliography of the Taylor movement and of the many ramifications of scientific management would fill a small volume. C. Bertrand Thompson's *Theory and Practice of Scientific Management* (Boston and New York, 1917) contains an extensive bibliography of the earlier literature; a select list of references on scientific management and the efficiency movement was prepared in 1913 by H. H. B. Meyer, chief bibliographer in the Library of Congress, and published in the series, *Special Libraries*, in May of that year; a bibliography of production engineering and cost accounting prepared by Paul M. Atkins and published by the Society of Industrial Engineers in 1927 is also useful. The early files of the *Bulletin of the Taylor Society* and its successor, the *Journal of the Society for the Advancement of Management* (later entitled *Advanced Management*) afford a convenient guide to the development of the movement, its philosophy, and its methods.

NOTES

Introduction

1. National Archives, Washington, D.C., War Records Division, Army Section, Ordnance Department (referred to hereafter as ODR), box 269, file 213/891: stenographic report of interview held on February 17 and 18, 1915, between Colonel C. B. Wheeler, R. F. Hoxie, R. G. Valentine, and J. P. Frey.

2. *Evidence Taken by the Interstate Commerce Commission in the Matter of Proposed Advances in Freight Rates by Carriers, August to December, 1910* (Senate Document 725, 61st Congress, 3rd session, Washington, D.C., 1911).

3. *Hearings Before the Special Committee of the House of Representatives to Investigate the Taylor and Other Systems of Shop Management* (3 vols., 62nd Congress, 2nd session, Washington, D.C., 1912; referred to hereafter as *Taylor System Hearings*).

4. Milton J. Nadworny, *Scientific Management and the Unions, 1900–1932: A Historical Analysis* (Cambridge, Mass., 1955), especially chapters 2–4.

5. Letter, John P. Frey, president emeritus, metal trades department, American Federation of Labor, to the author, January 28, 1953.

6. Two executive orders signed by President Theodore Roosevelt (no. 163, dated January 31, 1902, and no. 402, dated January 25, 1906) prohibited employees serving in government departments from soliciting increases in pay or attempting to influence legislation except through the head of the department concerned, on penalty of dismissal. These orders, however, were not invoked, nor was any reference made to them, during the agitation against scientific management in government work.

CHAPTER 1. The Taylor System

1. Of the great number of books and articles dealing with Taylor and his system of management, the following are particularly noteworthy: Frank B. Copley, *Frederick W. Taylor, Father of Scientific Management* (2 vols., New York, 1923); C. Bertrand Thompson, *The Theory and Practice of Scientific Management* (Boston, New York, and Chicago, 1917); Robert F. Hoxie, *Scientific Management and Labor* (New York and London, 1915); Horace B. Drury, *Scientific Management: A History and Criticism* (New York, 1915); Jean T.

McKelvey, *AFL Attitudes toward Production, 1900–1932* (Ithaca, N.Y., 1952); and Milton J. Nadworny, *Scientific Management and the Unions, 1900–1932: A Historical Analysis* (Cambridge, Mass., 1955).

2. ODR, Watertown Arsenal document file, box 266: Wheeler's replies to Hoxie's questionnaire.

3. Dartmouth College Conferences, Tuck School Conference, *Scientific Management* (Hanover, N.H., 1912), pp. 174–175.

4. ODR, Watertown Arsenal document file, box 269, file 213/904: memorandum by Dwight Merrick on "Time Allowance Percentage for Fatigue"; *ibid.*, file 10206–10222: blueprint of an instruction card used at Watertown Arsenal, memorandum on reverse.

5. Frederick W. Taylor, "Notes on Belting," *Transactions of the American Society of Mechanical Engineers, 1893;* "On the Art of Cutting Metals," *ibid., 1906.* See also Copley, *Taylor,* I, 243–244.

6. Frederick W. Taylor, *The Principles of Scientific Management* (New York and London, 1911); Taylor, *Shop Management* (New York, 1911); Harlow S. Person, *Graphical Analysis of Scientific Management* (New York, 1944); C. B. Thompson (ed.), *Scientific Management* (Cambridge, Mass., 1914); Thompson, "Scientific Management in Practice," *Quarterly Journal of Economics,* XIX (February 1915), 262–307.

7. See Taylor, "On the Art of Cutting Metals"; Copley, *Taylor,* I, 429–444 and II, 79–118. The account which follows rests wholly on these sources.

8. In most cases the cutting speeds of the machine tools were set by the makers and by the systems of pulleys, belts, and countershafts in the shop. Even if the machinist knew the correct cutting speed, he could make no use of his knowledge without putting new pulleys on the countershafts of his machine and probably making changes in the shape and heat treatment of his tools as well. See Taylor, *Principles,* pp. 112–113.

9. *Taylor System Hearings,* pp. 1483–1484, testimony of F. W. Taylor.

10. ODR, Watertown Arsenal document file, box 1053, file 10222: James Mapes Dodge to General William Crozier, April 23, 1909.

11. Taylor listed twelve variables which together determined the correct speed and feed of the machine: the hardness of the metal, the chemical composition and heat treatment of the tool, the thickness of the shaving cut by the tool, the contour of the cutting edge, the cooling method used, the depth of the cut, the duration of the cut, the lip and clearance angles of the tool, the elasticity of the work and the tool (*i.e.,* the amount of tool chattering), the diameter of the casting or forging being cut, the pressure of the chip or shaving on the cutting surface of the tool, and the pulling power of the machine (*Principles,* pp. 107–109). In practice these reduce to three principal

variables: the type of tool, the peripheral speed of the casting, and the nature of the metal being cut.

12. Daniel Bell, *Work and Its Discontents: The Cult of Efficiency in America* (Boston, 1956), pp. 3 ff.

13. *Transactions, A.S.M.E., 1886*, VII, 428–432.

14. On the general problems of work measurement and incentive-payments schemes, see Adam Abruzzi, *Work Measurement: New Principles and Procedures* (New York, 1952); William Gomberg, *A Trade Union Analysis of Time Study* (Chicago, Science Research Associates, 1948); R. S. Uhrbrock, *A Psychologist Looks at Wage-Incentive Methods* (New York, American Management Association, 1935: Institute of Management Series, no. 15); T. N. Whitehead, *The Industrial Worker* (2 vols., Cambridge, Mass., 1938).

15. Drury, *Scientific Management*, pp. 30–52.

16. Henry R. Towne, "Gain-Sharing," *Transactions, A.S.M.E., 1889*, vol. X. Towne's plan was put into operation in the Yale & Towne Manufacturing Company in 1887; after some initial success it was abandoned in the 1890's.

17. Drury, *Scientific Management*, pp. 39–41.

18. Frederick A. Halsey, "A Premium Plan of Paying for Labor," *Transactions, A.S.M.E., 1891*, XII; Halsey, "The Premium Plan of Paying for Labor," in John R. Commons (ed.), *Trade Unionism and Labor Problems* (New York, 1905), pp. 274–288. Halsey's plan, which was later adopted at Watertown Arsenal, is still one of the standard incentive-payments schemes.

19. Frederick W. Taylor, "A Piece-Rate System, Being a Step Toward Partial Solution of the Labor Problem," *Transactions, A.S.M.E., 1895*, XVI. Drury asserts (*Scientific Management*, p. 54, footnote) that Taylor's plan, though announced later, actually antedated the schemes of Towne and Halsey. Taylor in 1895 said his plan had been in effect at Midvale for ten years. Halsey in 1899 asserted that his plan was about fifteen years old. Probably Taylor, Towne, and Halsey were independently wrestling with the same problem at about the same time.

20. Taylor, "Piece-Rate System," p. 887: "With the differential rate the price will, in nine cases out of ten, be much lower than would be paid per piece either under the ordinary piece-work plan or on day's work."

21. See Thompson, *Theory and Practice*, pp. 146–147. If Taylor had admitted that the basic wage could *not* be determined by "scientific management," he could hardly have avoided admitting that it was a possible matter for collective bargaining. This he would not grant.

22. For a discussion of the inherent weaknesses of systems of payments by results, see the review of N. C. Hunt, *Methods of Wage*

Payment in British Industry (London and New York, 1951), by William B. Wolf in *American Economic Review*, XLIII, no. 1 (March 1953), 213–216. Wolf emphasizes that incentive-payments schemes may actually sharpen tensions between management and labor by translating into monetary units differences of opinion as to appropriate standards of output and proper allowances for variations in job conditions. This is in addition to the difficulty of establishing precise standards of performance and the inevitable distortion of traditional wage structures.

23. Sumner H. Slichter, *Union Policies and Industrial Management* (Washington, D.C., 1941), p. 283.

24. On this general subject, see Slichter, *Union Policies*, chapters X and XI; Lloyd Ulman, *The Rise of the National Trade Union: The Development and Significance of Its Structure, Governing Institutions, and Economic Policies* (Cambridge, Mass., 1955), chapter XV; and Van Dusen Kennedy, *Union Policy and Incentive Wage Methods* (New York, 1945), chapters II and III.

25. Taylor was prepared to admit that trade unions, particularly in England, had rendered useful service in shortening the hours of work and improving working conditions ("A Piece-Rate System," p. 882). As a general statement of his attitude, however, the text is correct. See Nadworny, *Management and the Unions*, pp. 20–22 and *passim*.

26. Taylor, "Why Manufacturers Dislike College Students," *Proceedings of the Society for the Promotion of Engineering Education* (1909), cited by Nadworny, *Management and the Unions*, p. 156, note 20.

27. Taylor Papers (Stevens Institute of Technology, Hoboken, N.J.), Crozier to Taylor, January 20, 1909; ODR, box 1053, file 10222: Taylor to Crozier, January 22, 1909.

28. Compare Frederick Taylor's statement: "Workingmen are not angels, but whatever else they are, they are not damned fools. All that is necessary is for a workman to have one object lesson of that sort [rate cutting] and he soldiers for the rest of his life." Cited in S. B. Mathewson, *Restriction of Output Among Unorganized Workers* (New York, 1931), p. 173.

29. See Everett C. Hughes, "The Sociological Study of Work," *American Journal of Sociology*, LVII, no. 5 (March 1952), 423–426; Wilbert E. Moore, "Current Issues in Industrial Sociology," *American Sociological Review*, XII, 651–657; Max Weber, "Zur Psychophysik der industriellen Arbeit," *Gesammelte Aufsaetze zur Soziologie und Sozialpolitik* (Tübingen, 1924), pp. 61–255. Weber was apparently the first to make the point, now generally accepted by industrial sociologists, that restriction of output can occur in the absence of unions, and without conscious agreement, wherever the work force

or even a considerable fraction of it feels some measure of solidarity. Hughes points out that restriction of output can and does take place even where there is no employer, as part of the general group process of determining levels of effort and product. See also Stanley B. Mathewson, *Restriction of Output Among Unorganized Workers* (New York, 1931), and *Trade Union Regulation and Restriction of Output*, Eleventh Special Report of the U.S. Commissioner of Labor, 1904 (Washington, 1904).

CHAPTER 2. *The Ordnance Department*

1. George W. Cullum, *Biographical Register of the Officers and Graduates of the U.S. Military Academy at West Point, N.Y., from Its Establishment, in 1802, to 1890* (3rd edition, Boston and New York, 1891). Crozier was fifth in rank in the class of 1876. See also "Crozier, William," *Encyclopedia Americana* (1956), VIII, 253.

2. William Crozier, *Ordnance and the World War: A Contribution to the History of American Preparedness* (New York, 1920), pp. 1–2, 22–24.

3. See, for example, the Annual Reports of the Chief of Ordnance, 1908, pp. 24–25; 1909, pp. 12–14; and 1910, pp. 585–586 (referred to hereafter as Annual Reports).

4. Crozier, *Ordnance and the World War*, pp. 22–23; *Taylor System Hearings*, p. 1119, testimony of General Crozier.

5. *Taylor System Hearings*, pp. 1120–1121, testimony of General Crozier.

6. Note, however, that there was probably a time lag between wage increases in private industry and parallel increases at the arsenals. It took time to gather data, hold hearings, and review existing wage levels.

7. ODR, box 269, file 213/891: material from Wheeler to accompany notes of Hoxie interview. For other data on labor turnover, see *ibid.*, box 266: replies to Hoxie's questionnaire. Corresponding estimates for private industry in this period are not readily available.

8. *Taylor System Hearings*, pp. 1112–1113, testimony of General Crozier.

9. *Ibid.*, p. 1878, testimony of Major Williams.

10. *Ibid.*, pp. 1112–1113, testimony of General Crozier; see also *ibid.*, p. 1120.

11. *Ibid.*, p. 1122.

12. *Ibid.*

13. Jeanette Mirsky and Allan Nevins, *The World of Eli Whitney* (New York, 1952), p. 268, footnote.

14. Crozier, *Ordnance and the World War*, p. 6.

15. *Ibid.*, pp. 7–8.

16. *Ibid.*, p. 8.

17. Fred H. Colvin, "Management at Watertown Arsenal," *American Machinist*, XXXVII, no. 11 (September 12, 1912), 424; Crozier, *Ordnance and the World War*, pp. 9–10.

18. Nadworny, *Management and the Unions*, pp. 65–66.

19. Crozier, *Ordnance and the World War*, pp. 5–6.

20. *Ibid.*, pp. 10–11.

21. ODR, box 1053, file 10222: Lieutenant Colonel Hobbs, Annual Report of principal operations at Watertown Arsenal during fiscal year ending June 30, 1907 (italics supplied).

22. Taylor Papers, Crozier to Taylor, January 9, 1909.

23. *Ibid.*, Crozier to Taylor, April 3, 1909.

24. *Ibid.*, Crozier to Taylor, March 14, 1911.

25. *Ibid.*, Crozier to Taylor, February 13, 1909.

26. *Ibid.*, Taylor to Barth, April 20, 1910. Earlier, in 1907, Taylor wrote to H. L. Gantt to tell him that there was a good chance of introducing the whole Taylor system in government shops. "There is no danger from strikes in the Government shops," he wrote, adding that with Theodore Roosevelt in Washington there would be "no serious interference." (*Ibid.*, Taylor to Gantt, January 3, 1907.)

27. *Ibid.*, Taylor to Crozier, March 17, 1911.

28. ODR, box 267, file 213/200–624: Crozier to Taylor, April 16, 1910, and Taylor to Crozier, April 20, 1910. Crozier wrote to inquire why, if the Taylor system was as good as Taylor said it was, it had led to a strike; or alternatively, if the Taylor system was no longer in use at Bethlehem, what had led the management to get rid of it. Taylor replied that his system was still in use in some of the departments at Bethlehem, and that in those departments the men were not discontented.

29. Nadworny, *Management and the Unions*, p. 31. See also ODR, box 1054, file 10222–10259, Crozier to Wilfred Lewis, April 15, 1909; box 1053, file 10222, Taylor to Crozier, April 15, 1909.

30. ODR, box 1053, file 10222: Hobbs to Crozier, February 8, 1909.

31. *Ibid.*

32. Annual Reports, 1907, 1908. The Annual Report, 1904, stated that high-speed steel had been used in the machine shops throughout the year, but that "the conditions existing in this shop are such that full advantage of this new steel cannot be secured."

33. Annual Report, 1908, p. 63.

34. ODR, Watertown Arsenal document file, box 266: endorsement by Wheeler dated February 2, 1909, on Thompson to Wheeler, November 9, 1908. See also *ibid.*, box 1053, file 10222: Wheeler to Crozier, February 2, 1909.

35. Taylor Papers, Wheeler to Crozier, January 25, 1909, enclosed in Taylor to Barth, January 29, 1909.

36. The very backward state of cost accounting at the arsenals is particularly remarkable in view of the fact that an officer of the Ordnance Department, Captain Henry Metcalfe, had done important pioneering work in costing methods. See his *The Cost of Manufactures and the Administration of Workshops, Public and Private* (New York, 1885) and "The Shop Order System of Accounts," *Transactions, A.S.M.E., 1886,* VII, 440–488. Taylor was fully conscious of his debt to Metcalfe and acknowledged it in his *Shop Management.*

37. ODR, Watertown Arsenal document file, box 266: endorsement by Wheeler on Thompson to Wheeler, November 9, 1908.

38. Taylor Papers, Wheeler to Crozier, January 25, 1909, enclosed in Taylor to Barth, January 29, 1909.

39. *Ibid.*

40. *Ibid.*

41. Copley, *Taylor,* II, 70–71; ODR, box 1053, file 10222: Hobbs to Crozier, February 8, 1909; Taylor Papers, Taylor to Crozier, December 17, 1906.

42. Taylor Papers, Taylor to Crozier, December 10, 1906.

43. *Ibid.,* Crozier to Taylor, December 14, 1906.

44. *Ibid.,* Crozier to Taylor, March 25, 1907.

45. *Ibid.,* Hobbs to Taylor, December 28, 1906.

46. *Ibid.,* Ruggles to Taylor, February 10, 1908.

47. *Ibid.,* Taylor to Ruggles, February 17, 1908.

48. *Ibid.,* Crozier to Taylor, January 25, 1909. See also *ibid.,* Crozier to Taylor, February 8, 1909.

49. *Ibid.,* Wheeler to Crozier, January 25, 1909.

50. ODR, box 1053, file 10222: Wheeler to Crozier, February 2, 1909.

51. Nadworny, *Management and the Unions,* p. 31; Taylor Papers, Gantt to Taylor, February 18 and March 17, 1908; Taylor to Gantt, March 21, 1908.

52. ODR, box 1053, file 10222: Taylor to Crozier, January 29, 1909.

53. Taylor Papers, Taylor to Barth, January 29, 1909.

54. Copley, *Taylor,* II, 23.

55. If for no other reasons, Barth has a certain claim to fame for his refusal to testify before the House Committee on the Taylor System in 1911 until the windows of the room were opened and some fresh air admitted.

56. ODR, Office of the Chief of Ordnance, correspondence file 230.437/380: Montgomery to Wheeler, January 19, 1918; see also Taylor Papers, Crozier to Taylor, July 22, 1913. Interpersonal friction between Barth and the officers was not unheard of at Watertown either, particularly after the molders' strike. See Taylor Papers, Barth to Taylor, August 28 and September 23, 1911.

57. *Taylor System Hearings,* pp. 588–589, testimony of F. W. Taylor.

CHAPTER 3. The Arsenal

1. ODR, Watertown Arsenal document file, box 266: replies to Hoxie's questionnaire.
2. Annual Reports, 1908–1913.
3. *Taylor System Hearings,* p. 498, testimony of Major Williams.
4. *Ibid.,* p. 411, testimony of Alexander Crawford.
5. *Ibid.,* p. 328, testimony of J. A. Mackean.
6. ODR, Watertown Arsenal document file, box 269, file 213/850–899: Wheeler's remarks at Hoxie interview. Asked whether it would have been desirable to have radically altered the layout of the arsenal before beginning managerial reforms, if money could have been obtained, Wheeler replied: "If we were a private industry, and had assurance . . . of continued work for a considerable period, I think it would have been desirable, but as we work entirely under appropriations made by Congress, and as these appropriations are, to a great extent, uncertain, I do not feel that we would have been justified in undertaking any great remodeling of the plant at this time."
7. *Taylor System Hearings,* pp. 200, 403, testimony of Mr. O'Leary.
8. *Ibid.,* pp. 1923–1924, testimony of Major Williams. For similar testimony from one of the molders, see *ibid.,* pp. 1896–1897, testimony of Mr. Kane.
9. ODR, Watertown Arsenal document file, box 266: replies to Hoxie's questionnaire, section entitled "Organization of Shop."
10. *Ibid.;* see also Fred H. Colvin, "Management at Watertown Arsenal," *American Machinist,* XXXVII, no. 11 (September 12, 1912), 424–428.
11. ODR, Watertown Arsenal document file, box 266: Crozier to Barth, February 6 and February 25, 1909; *ibid.,* box 1053, file 10222: Taylor to Crozier, February 8, 1909.
12. *Ibid.,* box 1052, file 10206–10222: Barth to Crozier, April 17, 1909.
13. Copley, *Taylor,* II, 176. For Taylor's indignant denial that his financial interest in the Tabor Company influence the purchase of equipment for use with the Taylor system, see *Taylor System Hearings,* p. 1504.
14. Taylor Papers, Taylor to Crozier, April 8, 1909.
15. Address given by Colonel Wheeler in Watertown Town Hall, May 1909, reprinted in *Taylor System Hearings,* pp. 108–114.
16. ODR, Watertown Arsenal document file, box 266: Barth to Wheeler, June 9, 1909.
17. *Ibid.,* box 1053, file 10222: Wheeler to Crozier, April 5, 1909, and copy of Executive Order No. 1079 (May 28, 1909) authorizing Barth's employment "for such a period as may enable him to make a

conclusive test of his methods in the operation of a manufacturing arsenal." Gantt received a fee of $100 a day when employed as consultant at Frankford Arsenal in 1911. Barth received a total of $9194 for 163 days' service between June 14, 1909 and November 30, 1911 (*Taylor System Hearings*, p. 1152).

18. *Ibid.*, box 267, file 400–499: Barth to Wheeler, October 2, 1911.

19. *Ibid.*, RG 156, Ordnance Department, Watertown Arsenal historical data, etc., box 1, 8E3 (marked "property of Major Williams"): Carl G. Barth, "A Broad Statement of the Main Things Attempted and Partly Accomplished by me, in my capacity as 'Expert in Shop Management' at the Watertown Arsenal, with some comments having a relation thereto."

20. *Ibid.*, Watertown Arsenal document file, box 266: "Outline of Chronology of Installation of the System of Scientific Management at the Watertown Arsenal up to the Time of Starting Time Studies" (to accompany p. 18 of Hoxie interview); Barth, "Broad Statement."

21. *Ibid.*, Office of the Chief of Ordnance, document file 10222–208–213: Wheeler to Crozier, November 13, 1909, December 2, 1909, and January 5, 1910.

22. *Ibid.*: Wheeler to Crozier, November 13, 1909. The actual cost was slightly smaller.

23. *Ibid.*: Wheeler to Crozier, January 5, 1910.

24. *Ibid.*: Wheeler to Crozier, December 2, 1909.

25. *Ibid.*: Wheeler to Crozier, February 16, 1910.

26. *Ibid.*, Watertown Arsenal document file, box 269, file 213/850–899: Wheeler's testimony at Hoxie interview.

27. *Taylor System Hearings*, p. 434, testimony of Major Williams.

28. ODR, Office of the Chief of Ordnance, document file 10222–208–213: Wheeler to Crozier, August 6, 1909, September 9, 1909, October 27, 1909, and January 5, 1910.

29. *Ibid.*, box 1054, file 10222–10259: Wheeler to Crozier, April 13, 1910.

30. *Ibid.*: Wheeler to Crozier, March 16, 1910; see also *Taylor System Hearings*, p. 498, testimony of Major Williams; ODR, Office of the Chief of Ordnance, document file 10222–208–213: Wheeler to Crozier, December 2, 1909.

31. *Ibid.*, box 1052, file 10206–10222: Barth to Crozier, April 17, 1909.

32. *Ibid.*, Office of the Chief of Ordnance, document file 10222–208–213: Wheeler to Crozier, September 9, 1909. The total cost of the belt maintenance equipment was $610; see *ibid.*, box 267, file 500–549: endorsement by Wheeler on Peirce to Wheeler, January 30, 1912.

33. *Ibid.*, Office of the Chief of Ordnance, document file 10222–208–13: Wheeler to Crozier, January 5, 1910.

34. *Ibid.*, box 1053, file 10222: Wheeler to Crozier, February 23, 1909.

35. *Ibid.*, Office of the Chief of Ordnance, document file 10222–208–213: Wheeler to Crozier, September 9, 1909.

36. *Ibid.*, Watertown Arsenal document file, box 269, file 213/850–899: Wheeler's testimony at Hoxie interview.

37. *Taylor System Hearings*, pp. 396–397, testimony of Colonel Wheeler.

38. ODR, box 1054, file 10222–10259: Wheeler to Crozier, April 13, 1910; box 1053, file 10222: Wheeler's report on operations at Watertown during year ending June 30, 1911.

39. *Ibid.*, Watertown Arsenal document file, box 269, file 213/850–899.

40. *Ibid.*: "Outline of Chronology"; Barth, "Brief Statement."

41. *Ibid.*: "Outline of Chronology"; box 1054, file 10222–10259: Wheeler to Crozier, April 13, 1910.

42. Annual Report, 1910, pp. 621–623; the cost of heat, light, and power at Watertown Arsenal increased from $20,599 in 1911 to $22,701 in 1912, $23,620 in 1913, and $29,038 in 1914 (ODR, box 269, file 213/891: material to accompany notes of Hoxie interview).

43. ODR, Watertown Arsenal document file, box 268, file 213/850–899: "Outline of Chronology."

44. *Ibid.*, box 267, file 250–299: Wheeler to Barth, August 22, 1910. Barth was also working during this period at the Franklin Manufacturing Company and the Forbes Lithograph Manufacturing Company.

45. *Taylor System Hearings*, pp. 85–86, testimony of Colonel Wheeler. Note that in this, as in other examples of the use of high-speed steel, the emphasis is entirely on the rate at which the metal could be cut, and not on accuracy. A skilled machinist working to close tolerances could not match the rate of cut in Wheeler's example, and would not try to. The advantages of Taylor tools seem to have been greatest for roughing work, rather than for finishing.

46. See, for example, the Annual Report, 1904, for the disappointing results secured from the initial introduction of high-speed steel at Watertown.

47. ODR, box 267, file 250–299; *Taylor System Hearings*, pp. 1159–1162, testimony of General Crozier.

48. *Taylor System Hearings*, p. 1088, testimony of Major King.

49. Annual Report, 1910, pp. 621–623; 1911, p. 54.

50. ODR, box 1054, file 10222–10259: Wheeler to Crozier, January 25, 1911.

51. Taylor Papers, Taylor to Crozier, January 12, 1911. See also *ibid.*, Taylor to Crozier, February 6, 1911, and Taylor to Barth, January 12, 1911.

52. ODR, box 267, file 300–399: Crozier to Secretary of War, February 13, 1911; *ibid.*: Crozier to Merrick, January 30, 1911; *Taylor System Hearings*, p. 1152, testimony of General Crozier.

53. *Taylor System Hearings,* p. 1181, testimony of General Crozier. The punctuation of this statement has been slightly corrected. The official transcript begins the second paragraph: "Now, we have another arbitrary rule, that for finishing the job within another time, which I am going to mention in a minute; we will pay the man for half the time he saves; and that half is arbitrary." Questioned about the arbitrary nature of the one-third rule, Crozier explained: "The object of that is to provide for inaccuracies, and to allow people to get used to the new methods. It has some consideration, however, for the slower man, and it is a matter that can very well be the subject for negotiation and collective bargaining." (*Ibid.,* pp. 1181–1182.)

54. *Ibid.,* p. 195, testimony of Colonel Wheeler.

55. *Ibid.,* pp. 156–157. The point was first brought out by John O'Leary of the International Molders' Union and was later confirmed by Colonel Wheeler.

56. Taylor, *Principles of Scientific Management,* pp. 130–131; R. F. Hoxie, *Scientific Management and Labor,* pp. 41–49; Drury, *Scientific Management,* pp. 69–71.

57. Taylor, *Principles,* pp. 131–132.

58. ODR, Watertown Arsenal document file, box 269, file 213/850–899: testimony of Colonel Wheeler at Hoxie interview, p. 72; *Taylor System Hearings,* pp. 104, 1864–1868.

59. *Taylor System Hearings,* pp. 82–83, testimony of Colonel Wheeler.

60. *Ibid.,* p. 74, testimony of Colonel Wheeler.

61. ODR, box 269, file 213/891: testimony of Colonel Wheeler at Hoxie interview.

62. *Taylor System Hearings,* pp. 417–418, testimony of Colonel Wheeler.

63. Major G. W. H. Tschappat, "Cost Keeping at Manufacturing Arsenals," *Iron Age,* March 28, 1935, pp. 14 ff.

64. *Taylor System Hearings,* p. 78, testimony of Colonel Wheeler.

65. ODR, box 627, file 400–499: Wheeler to Hon. W. B. Wilson, November 8, 1911.

66. See, for example, A. Hamilton Church, *The Proper Distribution of Expense Burden* (New York, 1921). On general developments in cost accounting in this period, see S. Paul Garner, *Evolution of Cost Accounting to 1925* (University, Alabama, 1954).

67. ODR, box 627, file 400–499: Wheeler to Hon. W. B. Wilson, November 8, 1911.

68. For a detailed description of the machine rate system as it was understood at that time, see Holden A. Evans, *Cost Keeping and Scientific Management* (New York, 1911). Evans was a naval constructor at Mare Island Navy Yard; Nadworny (*Management and the Unions,* p. 29) refers to him as "the unofficial Taylor representative on the West Coast."

69. *Taylor System Hearings*, pp. 791 ff., testimony of Colonel Wheeler.

70. *Ibid.*, p. 74 and pp. 82–83, testimony of Colonel Wheeler.

71. *Ibid.*, pp. 1081–1083, testimony of Major King. The system used at Rock Island Arsenal, to which King was attached, was modeled on that used at Watertown.

72. *Ibid.*, pp. 115 ff., testimony of Colonel Wheeler; ODR, Watertown Arsenal document file, box 266: replies to Hoxie's questionnaire.

73. *Taylor System Hearings*, pp. 82–83, testimony of Colonel Wheeler.

74. ODR, Watertown Arsenal document file, box 266: replies to Hoxie's questionnaire.

75. *Taylor System Hearings*, pp. 82–83, testimony of Colonel Wheeler.

76. *Ibid.*, p. 118, testimony of Colonel Wheeler.

77. *Ibid.*

78. The analysis that follows is based principally on the detailed description by Major Williams in *Taylor System Hearings*, pp. 459–465, but see also the testimony of Colonel Wheeler in the same source, particularly pp. 75, 82–83, 115–116, and 417–418.

79. *Ibid.*, p. 75, testimony of Colonel Wheeler.

80. ODR, Watertown Arsenal document file, box 266: replies to Hoxie's questionnaire.

81. *Ibid.*, box 269, file 213/891: Wheeler's testimony at Hoxie interview.

82. Compare *Taylor System Hearings*, pp. 321–322, testimony of J. A. Mackean, machine-shop foreman. Mackean pointed out that under the old system the foremen had many more written instructions from "the office" than under the Taylor system.

83. *Ibid.*, pp. 1153–1154, testimony of John O'Leary and General Crozier.

84. *Ibid.*, pp. 315–316, testimony of J. A. Mackean.

CHAPTER 4. Conflicts

1. *Taylor System Hearings*, pp. 1864–1868, memorandum from the commanding officer, Watertown Arsenal, on the strike of the molders. See also Taylor Papers, Taylor to M. L. Cooke, June 6, 1911, informing Cooke that the first "task" had been set a few days earlier at Watertown on gear-cutting machines, that the worker had come within 10 per cent of earning his bonus, and that Taylor "trusted" that he had by this time earned it.

2. ODR, box 267, file 300–399: Wheeler to Robert H. Rice, March 1, 1911; to Dr. Bradley Stoughton, March 1, 1911; to George W. Cope, editor, *Iron Age*, February 20 and 25, 1911; and Cope to Wheeler, February 23, 1911.

3. Taylor Papers, Taylor to Crozier, June 9, 1911.

4. *Ibid.*, Crozier to Taylor, June 23, 1911, quoting Taylor's remarks while in Washington.

5. *Ibid.*, Crozier to Taylor, June 23, 1911.

6. *Ibid.*, Taylor to Crozier, June 26, 1911.

7. *Ibid.*

8. *Ibid.*, Barth to Taylor, August 12, 1911.

9. *Taylor System Hearings,* p. 1891, testimony of Joseph R. Cooney, molder.

10. For Wheeler's and Crozier's explanations, see *Taylor System Hearings,* p. 1875.

11. See above, p. 72.

12. *Taylor System Hearings,* p. 167, testimony of Colonel Wheeler.

13. *Ibid.*, pp. 1864–1868, memorandum from the commanding officer, Watertown Arsenal, on the strike of the molders.

14. *Ibid.*

15. *Ibid.*, p. 197, testimony of Colonel Wheeler.

16. *Ibid.*, pp. 1923–1924, testimony of Major Williams.

17. *Ibid.*

18. *Ibid.*, pp. 1864–1868, memorandum from the commanding officer, Watertown Arsenal, on the strike of the molders.

19. *Ibid.*

20. *Ibid.*, p. 1916, testimony of Dwight Merrick.

21. Detailed descriptions of the events in the foundry on August 10 and 11 are to be found in *Taylor System Hearings,* pp. 227 ff. and 1888–1889 (testimony of J. R. Cooney); *ibid.*, pp. 1871 ff. (Lieutenant Colonel Thompson's report on the strike); ODR, file 10206–10222, box 1052 (Colonel Wheeler's Annual Report, 1912); and "Scientific Management at United States Arsenals: Results Accomplished at Watertown," *Iron Age,* LXXXVIII (November 9, 1911), 1022–1024.

22. *Taylor System Hearings,* p. 1930, testimony of Major Williams.

23. *Ibid.*, pp. 1930–1931, testimony of Dwight Merrick.

24. On the circumstances surrounding this meeting, see *Taylor System Hearings,* p. 230, testimony of Joseph R. Cooney.

25. *Ibid.*, pp. 1888–1889, testimony of Joseph R. Cooney.

26. *Ibid.*

27. *Ibid.*

28. *Ibid.*, pp. 1864–1868, memorandum from the commanding officer, Watertown Arsenal, on the strike of the molders.

29. *Ibid.*

30. *Ibid.*, p. 1886.

31. *Ibid.*, pp. 227 ff., testimony of Joseph R. Cooney.

32. *Ibid.*, pp. 1864–1868, memorandum from the commanding officer, Watertown Arsenal, on the strike of the molders.

33. A.F.L. Archives, molders file (1911), Watertown Arsenal molders to Massachusetts Senators and Congressmen, August 11, 1911.

34. A.F.L. Archives, molders file (1911), William John (secretary, I.M.U. no. 106) to Frank M. Morrison, August 14, 1911.

35. Letter, John P. Frey to the author, January 28, 1953.

36. *Ibid.*

37. Frank T. Stockton, *The International Molders Union of North America* (Baltimore, 1921), pp. 157–158.

38. The *International Molders' Journal,* reporting the strike and the board's decision, glossed over the fact that official approval had been retroactive, stating with deceptive candor that "Ever since the 'system' was introduced there has been friction, and finding that all efforts to secure an understanding with the commandant of the arsenal met with failure the matter was laid before our Executive Board in the same manner that all other grievances are presented," XLVII, no. 9 (September 1911).

39. *Taylor System Hearings,* p. 65, testimony of Colonel Wheeler.

40. Taylor Papers, Barth to Taylor, August 12, 1911.

41. *Taylor System Hearings,* p. 1886, Barth's testimony at Thompson inquiry.

42. Taylor Papers, Barth to Taylor, August 15, 1911.

43. *Ibid.,* Barth to Taylor, August 28, 1911.

44. *Ibid.,* Barth to Taylor, September 23, 1911.

45. *Ibid.,* Crozier to Taylor, September 16, 1911.

46. *Taylor System Hearings,* p. 1885, Barth's testimony at Lieutenant Colonel Thompson's inquiry.

47. *Industrial Relations Commission Hearings,* p. 890, testimony of Carl Barth.

48. See, for example, Hoxie, *Scientific Management and Labor,* pp. 39–61, and John P. Frey, "Labor's Attitude Towards Scientific Management," in Daniel Bloomfield (ed.), *Modern Industrial Movements* (New York, 1920), p. 151.

49. For Gantt's views see *Taylor System Hearings,* pp. 579–580: "A man who did not know anything more than what he can get from a stop watch would be absolutely useless . . . to send a clerk out into the shop with a stop watch to study anything is perfectly absurd. You want to take the man who has the most knowledge of the subject to start with, and give him a chance to study the details . . . unless the man who is making this study can command the respect of the men he is not worth shucks."

50. Taylor Papers, Taylor to Cooke, August 20, 1911.

51. *Ibid.,* Taylor to Barth, August 14, 1911.

52. *Ibid.,* Taylor to Crozier, September 21, 1911.

53. *Ibid.,* Taylor to Crozier, July 5, 1912.

54. *Ibid.,* Taylor to Crozier, September 21, 1911.

55. *Ibid.,* Taylor to Wheeler, February 9, 1914.

56. *Taylor System Hearings,* pp. 1500–1501, testimony of F. W. Taylor.

57. Taylor Papers, Taylor to Barth, August 4, 1911.

58. *Taylor System Hearings*, p. 195, testimony of Major Williams.

59. *Ibid.*, p. 105, testimony of Colonel Wheeler.

60. ODR, box 269, file 213/891, p. 74: Wheeler's testimony at Hoxie interview.

61. *Taylor System Hearings*, pp. 429–430, testimony of Colonel Wheeler and E. H. Fitzgerald.

62. *Ibid.*, p. 1165, testimony of General Crozier.

63. *Ibid.*, pp. 21–22, testimony of Mr. Jennings.

64. *Ibid.*, pp. 520–521, testimony of Mr. Jennings.

65. *Ibid.*, pp. 1864–1868, memorandum from the commanding officer, Watertown Arsenal, on the strike of the molders.

66. Edgecomb was employed at the arsenal from August 19 to September 20, 1911. For his performance and the molders' reactions, see *Taylor System Hearings*, pp. 66 ff., 131, 140, 144, 151, 202–203, and 1864–1868. Major Williams stated that Edgecomb's molds were indistinguishable from the others, that the molders in any case did too much finishing of their molds, and that if Edgecomb received unusual privileges, it was from the foreman and not by order of the officers. He also claimed that, as a result of Edgecomb's example, one of the regular molders increased his production rate considerably—an illustration, if true, of the disruptive effects of the rate buster on the solidarity of the group.

67. *Ibid.*, p. 1221, testimony of General Crozier.

68. *Ibid.*, pp. 1223–1224.

69. Stockton, *International Molders Union*, pp. 157–158.

70. *Machinists Monthly Journal*, XVII (October 1905), 922, cited Nadworny, *Management and the Unions*, p. 25.

71. In strict accuracy, the introduction of certain Taylor methods in the Navy yards antedated Hobbs's experiments at Rock Island, though they did not provoke the same violent reaction. Holden Evans, working at the Mare Island Navy Yard in 1907–1908, introduced a piece rate system for ship scalers that seems to have been based on time study and that certainly stemmed from Taylor's ideas. Admiral Casper F. Goodrich, a friend of Taylor's since the 1890's, was instrumental in reforming the Navy's system of ordering tool steels, and both Hathaway and Barth were hired as consultants at Navy yards on the East coast during 1908. None of these episodes represented the adoption of the Taylor system as a whole. Except perhaps in Evans' work, the stop watch was not used, and the innovations were not formally and publicly identified as "the Taylor system." As evidence that they were not known as the Taylor system, see *Taylor System Hearings*, p. 1350, testimony of Charles A. Osborne (machinist at Norfolk Navy Yard): "We had as many different methods in the yard, and have now, as a dog is full of fleas."

72. ODR, box 1053, file 10222: Hobbs to Crozier, February 8, 1909.

73. Nadworny, *Management and the Unions*, p. 51.

74. Samuel Gompers, "The Miracles of 'Efficiency,'" *American Federationist*, XVIII, no. 4 (April 1911), 273–279; "Machinery to Perfect the Living Machine," *ibid.*, XVIII, no. 2 (February 1911), 116–117, cited Nadworny, *Management and the Unions*, p. 51. Nadworny points out (pp. 53–54) that Gompers, in his private correspondence, evinced no doubt that scientific management was intended to and would destroy the unions, but that the articles in the *American Federationist* emphasized the speed-up rather than the threat to collective bargaining.

75. *Official Circular No. 12*, International Association of Machinists, office of the international president, Washington, D.C., April 26, 1911 (reprinted in *Taylor System Hearings*, pp. 1222–1223).

76. *Iron Age*, LXXXVIII (November 9, 1911), 1024.

77. *Hearings Before the Committee on Labor of the House of Representatives on H.R. 90*, *passim* (62nd Congress, 1st session, Washington, D.C., 1911).

78. See Taylor Papers, Taylor to Gantt, March 27, 1911: "When I was in Washington last I was told that Congressman Weeks, who is in the Watertown district, had stated that he, in three or four months past, had not received a single letter of complaint from workmen in the Watertown Arsenal, whereas when Barth first went there and before anything whatever was done, he was fairly inundated with letters."

79. "Scientific Management," *International Molders' Journal*, XLVII, no. 4 (April 1911); "Taylor's Scientific Shop Management," *ibid.*, XLVII, no. 6 (June 1911).

80. Taylor Papers, Taylor to Crozier, February 18, 1912.

81. *Ibid.*, Crozier to Taylor, October 10, 1912.

82. *Ibid.*, Taylor to Wheeler, January 28, 1914.

83. *Taylor System Hearings*, p. 230.

84. Letter, John P. Frey to the author, January 28, 1953.

85. *Taylor System Hearings*, p. 428, testimony of Alexander Crawford.

86. *Ibid.*, pp. 1888–1889, testimony of Joseph R. Cooney.

87. *Ibid.*, pp. 1153–1154, testimony of General Crozier. The molders alleged that Larkin, because of the additional duties imposed on him by the Taylor system, had asked for the assistance of a clerk. When this was refused he resigned, but his successor, Roach, was given an assistant. Compare the testimony of O'Leary (*ibid.* and pp. 1201–1202). Larkin had been employed at the arsenal for more than fourteen years.

88. *Ibid.*, p. 200, testimony of O'Leary.

89. *Ibid.*, p. 403, testimony of Wheeler and O'Leary.

90. Letter, Carl Huhndorff, director of research, International Association of Machinists, to the author, September 20, 1954.

91. *Ibid.*

92. *Machinists' Journal* (July 1911), p. 689.

93. A.F.L. Archives, I.A.M. file (1913), W. H. Johnston to Frank Morrison, February 17, 1913, reporting that the I.A.M. had succeeded in initiating about three hundred members at Watertown.

CHAPTER 5. *Consequences*

1. ODR, Watertown Arsenal document file, box 266: replies to Hoxie's questionnaire. (All percentages have been rounded to the nearest whole number.)

2. *Ibid.*

3. *Ibid.*

4. *Ibid.*

5. *Ibid.*, box 268, file 213/659. It should be noted that a number of different metals were cast (lead, aluminum, white metal, iron, steel and bronze) and that the proportion of the various metals in each month's output may have varied.

6. *Ibid.*, box 266: replies to Hoxie's questionnaire; *ibid.*, box 265, file 850–899: report of Hoxie interview.

7. *Taylor System Hearings*, p. 801, testimony of Colonel Wheeler, Exhibit K. Expenditures on power plant were $9852.02 in 1909, $16,-071.81 in 1910, and $17,300.00 in 1911 (fiscal years).

8. Cf. A. Hamilton Church, *Production Factors in Cost Accounting and Works Management* (New York, 1910), p. 11.

9. *Taylor System Hearings*, p. 792, testimony of Colonel Wheeler.

10. *Ibid.*, p. 74, testimony of Colonel Wheeler. For slightly different figures on the same product, see *ibid.*, p. 1152, testimony of General Crozier.

11. *Ibid.*, p. 489, testimony of Major Williams.

12. ODR, Watertown Arsenal document file, box 269, file 213/850–899: testimony of Colonel Wheeler at Hoxie interview, p. 35.

13. *Ibid.*: testimony of Colonel Wheeler at Hoxie interview.

14. *Ibid.*: box 267, file 213/550–599: Wheeler to Colvin, May 15, 1912. This figure does not include the cost of hiring Barth and Merrick. The total cost of installing (not maintaining) the Taylor system, up to May 1, 1912, was calculated as follows:

Cost of experts	$15,478.33
Making changes in machine tools	17,874.34
Construction of planning room	2,510.45
Equipment of planning room	673.94
Tool-grinding and belt-maintenance equipment	2,251.95
Storeroom and toolroom equipment	5,423.63
Stock of high-speed tool steels	5,000.00
	$49,212.64

15. *Taylor System Hearings*, p. 82, testimony of Colonel Wheeler.

16. ODR, Watertown Arsenal document file, box 269, file 213/891: material from Wheeler to accompany notes of Hoxie interview.

17. *Taylor System Hearings*, pp. 1174, 1198, 1204, 1210, testimony of General Crozier.

18. *Ibid.*, p. 1210, testimony of General Crozier.

19. *Ibid.*, p. 1204, testimony of O'Leary. This job was one of those done by Edgecomb the Canadian "rate-buster."

20. ODR, box 269, file 213/891: Wheeler's testimony at Hoxie interview.

21. *Ibid.*, box 266: Taylor to Crozier, April 8, 1909.

22. Taylor Papers, Crozier to Taylor, June 26, 1912.

23. ODR, file 10206–10222: Wheeler to Fred Colvin, April 19, 1912.

24. Taylor Papers, Crozier to Taylor, October 10, 1912.

25. *Taylor System Hearings*, p. 492.

26. *Ibid.*, p. 1570, testimony of Carl Barth.

27. Taylor Papers, Crozier to Taylor, February 10, 1912.

28. *Taylor System Hearings*, p. 363, testimony of Willard Barker.

29. *Ibid.*, p. 508, testimony of Olaf Nelson.

30. *Ibid.*, pp. 205–206, testimony of Gustave Lawson.

31. *Ibid.*, p. 35, testimony of Orrin Cheney.

32. *Ibid.*, p. 176, testimony of Joseph Hicklin.

33. *Ibid.*, pp. 430–436, testimony of James D. Reagan.

34. ODR, box 268, file 213/698: comments by Major Williams on an examination of a sample of forty-five out of three hundred and forty-nine employees of Watertown Arsenal who petitioned the Secretary of War, July 21, 1913, against continuation of the Taylor system.

35. *Taylor System Hearings*, p. 279, testimony of E. M. Burns.

36. *Ibid.*, p. 406, testimony of Alexander Crawford.

37. *Ibid.*, p. 215, testimony of Isaac Goostray.

38. *Ibid.*, p. 319, testimony of James A. Mackean.

39. *Ibid.*, p. 337.

40. *Ibid.*, p. 509, testimony of Olaf Nelson.

41. *Ibid.*, pp. 1890–1891, testimony of J. R. Cooney.

42. *Ibid.*, p. 1893, testimony of E. L. Sherman.

43. *Ibid.*, p. 176, testimony of Joseph Hicklin.

44. *Ibid.*, p. 238, testimony of J. R. Cooney.

45. *Ibid.*, p. 1902, testimony of George V. Miller.

46. *Ibid.*, p. 177, testimony of Joseph Hicklin.

47. *Ibid.*, p. 509, testimony of Olaf Nelson.

48. *Ibid.*, pp. 205–206, testimony of Gustave Lawson.

49. ODR, box 268, file 213/744: Colonel Gibson to Colonel Wheeler, November 20, 1913, forwarding transcript of evidence taken from foremen at Watervliet Arsenal.

50. *Taylor System Hearings,* p. 1904, testimony of Joseph Hicklin.
51. *Ibid.,* p. 1889, testimony of Gustave Lawson.
52. *Ibid.,* p. 1908, testimony of Martin Roach.
53. *Ibid.,* pp. 1896–1897, testimony of Thomas Kane.
54. *Ibid.,* p. 144, testimony of Joseph Hicklin.
55. *Ibid.,* p. 299, testimony of Richard H. Stackhouse.
56. *Ibid.,* p. 327, testimony of J. A. Mackean.
57. *Ibid.,* pp. 315–316.
58. *Ibid.*
59. *Ibid.,* pp. 507 ff., testimony of Olaf Nelson.
60. *Ibid.,* p. 406, testimony of Alexander Crawford.
61. *Ibid.,* pp. 508–510, testimony of Olaf Nelson.
62. *Ibid.,* p. 339, testimony of J. A. Mackean.
63. *Ibid.,* pp. 315–316.
64. ODR, box 268, file 213/698, enclosure 3.
65. Miner Chipman, "Efficiency, Scientific Management, and Organized Labor," a paper read at the annual meeting of the Efficiency Society, January 21, 1916.
66. Taylor Papers, Taylor to Barth, March 19, 1914.
67. *Ibid.,* Taylor to Crozier, February 2, 1914.
68. ODR, box 268, file 213/757: Crozier to Chipman, January 5 and December 2, 1915.
69. Chipman, "Efficiency, Management, and Labor."
70. *Taylor System Hearings,* p. 1158, testimony of General Crozier.
71. *Ibid.,* pp. 1223–1224, testimony of General Crozier.
72. Composed of William Wilson, William Redfield, and John Tilson.
73. The report praised the mechanical and administrative aspects of the Taylor system, such as the standardization of parts and the routing of components, but expressed misgivings that the system might, under certain employers, be used to the disadvantage of the workers. It also criticized the use of the stop watch for task setting and suggested that the worker should have the right to approve or disapprove time studies and piece rates. See Nadworny, *Management and the Unions,* pp. 63–64.
74. McKelvey, *AFL Attitudes toward Production,* p. 18.
75. Taylor Papers, Taylor to Barth, August 30, 1911.
76. *Ibid.,* Taylor to Crozier, July 5, 1912.
77. *Ibid.,* Taylor to Crozier, July 26, 1912.
78. For the contrasting situation in the Navy Department and the role of George von L. Meyer, Secretary of the Navy under President Taft, see Nadworny, *Management and the Unions,* pp. 65–67.
79. Taylor Papers, Crozier to Taylor, December 26, 1912.
80. *The Stop Watch and Bonus System in Government Work,* Hearing on H.R. 8662 (63rd Congress, 2nd session, Washington, D.C., 1914). Cf. Nadworny, *Management and the Unions,* pp. 82–83.

81. Taylor Papers, Taylor to Crozier, October 1, 1913.

82. *Ibid.*, Crozier to Taylor, October 6, 1913.

83. *Ibid.*

84. Nadworny, *Management and the Unions,* p. 83.

85. Taylor Papers, Crozier to Taylor, February 25, 1915.

86. *Ibid.,* Crozier to Taylor, March 4, 1915.

87. United States Statutes at Large, 38:1083.

88. At Frankford Arsenal, where the work was normally financed under the Army Bill, premiums had to be stopped. Both the Judge Advocate General of the Army and the Comptroller of the Treasury, however, were of the opinion that the legislation did not apply to piecework. The restrictions could be evaded, therefore, by a change in the form of the incentive-payments system. See ODR, box 269, file 213/907: Comptroller of the Treasury to Chief of Ordnance, March 22, 1915.

89. Taylor Papers, Crozier to Taylor, March 18, 1915; ODR, box 269, file 213/970: Crozier to Hon. James A. Gallivan, November 20, 1915.

90. Nadworny, *Management and the Unions,* p. 103.

91. In technical terms, the hypothesis is that the minimum point on the arsenal's average cost curve had been shifted down and to the right; or alternatively that the arsenal had been enabled to move down and to the right on its long-run "envelope" cost curve.

92. See, in particular, Nadworny, *Management and the Unions,* and McKelvey, *AFL Attitudes toward Production.*

93. *Bulletin of the Taylor Society,* March 1917, pp. 10–11; cited by McKelvey, *AFL Attitudes toward Production,* p. 26.

94. L. P. Alford, "An Industrial Achievement of the War," *Industrial Management,* LV (February 1918), 97–100.

95. ODR, Watertown Arsenal document file, box 266: replies to Hoxie's questionnaire; National Archives, Microfilm Publications, microcopy no. T-4, roll no. 13: William O. Thompson, *Report of the Commission on Industrial Relations on Organization of Government Employees and the Establishment of an Industrial Court at the Watertown Arsenal* (January 29, 1915).

96. Nadworny, *Management and the Unions,* p. 105.

INDEX

LIBRARY OF CONGRESS CATALOGING IN PUBLICATION DATA

Aitken, Hugh G. J.
Scientific management in action.

Reprint. Originally published: Taylorism at Watertown Arsenal.
Cambridge, Mass. : Harvard University Press, 1960.
With new foreword.

Bibliography: p.   Includes index.
Watertown Arsenal (Mass.) 2. Industrial management—Massachu-
setts—Case studies. 3. Taylor, Frederick Winslow, 1856–1915.
I. Title.

UF543.W4A55 1985   623.4'0685   84–26462
ISBN 0–691–04241–1 (alk. paper)
ISBN 0–691–00375–0 (pbk. : alk. paper)